Sewage - Recent Advances, New Perspectives and Applications

Edited by Tao Zhang

Published in London, United Kingdom

IntechOpen

Supporting open minds since 2005

Sewage - Recent Advances, New Perspectives and Applications
http://dx.doi.org/10.5772/intechopen.95668
Edited by Tao Zhang

Contributors
Katsuyuki Takahashi, Naoya Satta, Koichi Takaki, Mohd Salim Mahtab, Izharul Haq Farooqi, Girum Ayalneh Tiruye, Sami Ullah Bhat, Umara Qayoom, Karima Elkarrach, Fatima Atia, Anass Omor, Omar Laidi, Saloua Biyada, Mohamed Benlmelih, Mohammed Merzouki, Petia Mijaylova Nacheva, Danilo Gamboa-Santana, Edson B. Estrada-Arriaga, Roberto Uribe, Jean Engohang-Ndong

First published in London, United Kingdom, 2022 by IntechOpen
IntechOpen is the global imprint of INTECHOPEN LIMITED, registered in England and Wales, registration number: 11086078, 5 Princes Gate Court, London, SW7 2QJ, United Kingdom
Printed in Croatia

British Library Cataloguing-in-Publication Data
A catalogue record for this book is available from the British Library

Additional hard and PDF copies can be obtained from orders@intechopen.com

Sewage - Recent Advances, New Perspectives and Applications
Edited by Tao Zhang
p. cm.
Print ISBN 978-1-83969-824-8
Online ISBN 978-1-83969-825-5
eBook (PDF) ISBN 978-1-83969-826-2

For EU product safety concerns:
IN TECH d.o.o., Prolaz Marije Krucifikse Kozulić 3, 51000 Rijeka, Croatia, info@intechopen.com or visit our website at intechopen.com.

Meet the editor

Dr. Tao Zhang is an Associate Professor and Ph.D. Supervisor at the College of Resources and Environmental Sciences, China Agricultural University, China. His academic background covers waste management, wastewater treatment, utilization of agricultural waste. He is awarded the Scientific Chinese - Outstanding Young Scientist Award, the Innovation Award for Industry-University-Research Cooperation of China, the Character Award - Invention and Entrepreneurship Award of China Association of Inventions. His H-index is 23 (Scopus) and he has published more than 50 papers in Chemical Engineering Journal, Water Research, Journal of Hazardous Materials, Green Chemistry, Renewable and Sustainable Energy Reviews, and so on. Amongst, 11 ESI Highly Cited Paper and 4 ESI Hot Paper. He has authorized more than 20 Chinese invention patents.

Contents

Preface

Water is the source of life and a clean body of water is the key to human survival. However, the world's water is being polluted by sewage, which, if not treated, will endanger human health and threaten the environment. If we want to live on Earth in a sustainable and healthy way, it is our duty to treat sewage. Although sewage treatment has a long history, it is becoming increasingly difficult due to rapid population growth and the increasing variety of sewage. This requires us to: first, we must master the current situation of sewage, from the generation, discharge, destination and its impact of sewage; second, innovative sewage treatment technology, the generation of sewage is inevitable, we can do is to better solve the problem, innovative sewage treatment technology is a shortcut.

We need to systematically understand the harm brought by sewage, first of all, the harm to the environment, the human living environment if destroyed, the impact will be unbearable. The existence of sewage will lead to a reduction in the number of organisms in the ecosystem or even extinction, resulting in a reduction in the value of various types of environmental resources, and ultimately the ecological balance is destroyed; second, the harm to production, water pollution, industrial water must be invested in more treatment costs, resulting in a waste of resources, energy, food industry water requirements are more stringent, water quality is not qualified, will bring production to a halt. Agricultural use of sewage, so that crop yield reduction, lower quality, and even human and animal victims, large areas of farmland suffer from pollution, reducing the quality of soil; third, the harm to the human body, if people drink polluted water, will cause acute and chronic poisoning, cancer, infectious diseases, and some other bizarre diseases, the sensory deterioration caused by polluted water, will cause inconvenience to human life, emotions are adversely affected.

There are numerous hazards of sewage, so methods of sewage treatment are of great concern. Traditional wastewater treatment methods can no longer meet the existing new types of wastewaters, so it is urgent to explore more novel wastewater treatment methods. In the context of the new crown epidemic, the pollution condition of water bodies by drugs is becoming more and more serious, and we must be more careful in treating these new wastewaters due to the special characteristics of drugs. New wastewater treatment methods are escalating around physical, chemical, and biological methods, such as wastewater treatment using microbial fuel cells, oxidizing electric furnaces for wastewater treatment, and fused physicochemical methods. This manuscript systematically describes the research progress on wastewater, treatment methods, etc., which hopefully can be enlightening to the readers.

Tao Zhang
China Agricultural University,
Beijing, China

Section 1

Physicochemical Technology for Sewage Treatment

Chapter 1

A Novel Wastewater Treatment Method Using Electrical Pulsed Discharge Plasma over a Water Surface

Katsuyuki Takahashi, Koichi Takaki and Naoya Satta

Abstract

Electrical pulsed discharge plasma produces various powerful oxidizing agents, such as hydroxyl radicals and ozone, which have high oxidation potential. These species play an important role in the decomposition of persistent organic compounds in wastewater. Because highly concentrated oxidants are directly produced inside the plasma, plasma realizes high-speed wastewater treatment without pretreatment of samples, such as pH adjustment. The pulsed discharge plasma generated over the water surface and inside bubbles is highlighted as a highly efficient method for plasma generation and radical supply into wastewater. In this paper, the physical and chemical properties of the discharge plasma generated over a water surface are described. The decomposition of persistent organic compounds dissolved in wastewater, such as 1,4-dioxane, formic acid, and dichloromethane, by plasma discharge is demonstrated, and their mechanisms are discussed. These persistent compounds, which have strong toxicity and stability, can be efficiently decomposed and removed quickly from solutions by plasma treatment. Furthermore, the treatment of nutrient solutions used in hydroponic systems for plant cultivation is also introduced as a novel application of plasma, and the effects of bacterial inactivation, decomposition of allelochemicals, and improvement in plant growth by plasma are demonstrated.

Keywords: plasma, wastewater, persistent compounds, advanced oxidation process, hydroxyl radical, oxidation, ozone

1. Introduction

Electrical pulsed discharge under water, an advanced oxidation process (AOP), can instantaneously produce chemically active species such as hydroxyl radicals and ozone in a nonthermal plasma [1–3]. It is well known that these species are powerful oxidizing agents that play an important role in degrading persistent organic chemical compounds and sterilizing bacteria in wastewater [4, 5]. Hydroxyl radicals can contribute to the decomposition of persistent organic pollutants, which cannot be decomposed by conventional methods because of their very high oxidation potential. Wastewater treatment using discharge plasma has several advantages over other AOPs. In the plasma system, active species are directly produced in plasma

IntechOpen

through various reactions initiated by high-energy electron impact. These active species are directly dissolved into water, and then oxidation reactions are induced. Because only electricity-supplied metal electrodes and a high voltage power supply are needed to generate plasma, pretreatment of water, such as pH adjustment is not required for active species production, unlike other AOPs such as Fenton and ozone/hydrogen peroxide methods [6]. Plasma can also be adapted for the treatment of water, which has ultraviolet (UV) light non-transmitting properties, which is advantageous over AOPs utilizing UV light, such as photocatalytic processes and UV-hydrogen peroxide methods. In the case of plasma discharge in contact with water, the types of active species produced in plasma can vary by atmospheric gas species. The production of active species can be localized near the plasma channel. Thus, water treatment using plasma discharge can be useful in various uses. Furthermore, because the density of active species in plasma is very high, it can also be used for the treatment of highly concentrated wastewater. These characteristics of discharge plasma can make the system simple, easy to use and easy to install anywhere. Owing to these advantages, the plasma technique has attracted attention as a promising method in various fields, such as industrial and environmental wastewater treatment [4, 6, 7] and material [8], agricultural [9, 10], and medical applications [11].

In this article, characteristics of discharge plasma generated by pulsed high voltages, chemical reactions induced by plasma, and treatment of persistent organic pollutants such as 1,4-dioxane and dichloromethane dissolved in water by plasma discharges are introduced. Agricultural applications of discharge plasma in contact with water surfaces are also mentioned as a novel application.

2. Generation of pulsed discharge plasma

Plasma is one of the states of matter and consists of high-density electrons and ions produced through ionization reactions of neutral gas atoms and molecules. Ionization is initiated by an impact of high-energy electrons accelerated by a high electric field; then, the electron and ion densities rapidly increase due to an electron avalanche phenomenon, and plasma forms [12]. To generate a high electric field, a sharpened metal electrode to which a high voltage is applied is generally used. Although various kinds of voltage waveforms, such as DC and AC, are employed, a pulsed high voltage generated by a pulsed power generator is used for highly efficient plasma generation, [13, 14] especially under water. Pulsed power is a special technology used to generate a pulse voltage that has a high voltage amplitude over several tens of kV, high power over several tens of kW, short pulses on the order of nanoseconds, and high pulse frequencies over several kHz, achieved with a compact system [15].

When a pulsed high voltage is applied to pointed electrodes immersed in water, discharge plasma is generated at the tip of the electrode. One of the formation mechanisms of the discharge plasma in water using a long pulse voltage on the order of several μs is considered as follows: In the first step, the discharge is initiated in a small bubble on the electrode surface, and then it propagates into the water, as shown in **Figure 1**. The current in the high electric field region causes Joule heating and vaporization of the liquid, forming bubbles [16–18]. The electron density in the plasma directly generated under water is very high, on the order of 10^{25} m^{-3} [19], which can produce very high density chemically active species such as hydroxyl radicals and contribute to high-speed wastewater treatment. The plasma discharge under water can also produce strong shock waves with pressure on the order of several GPa. The problem with this method is that wastewaters exhibit high

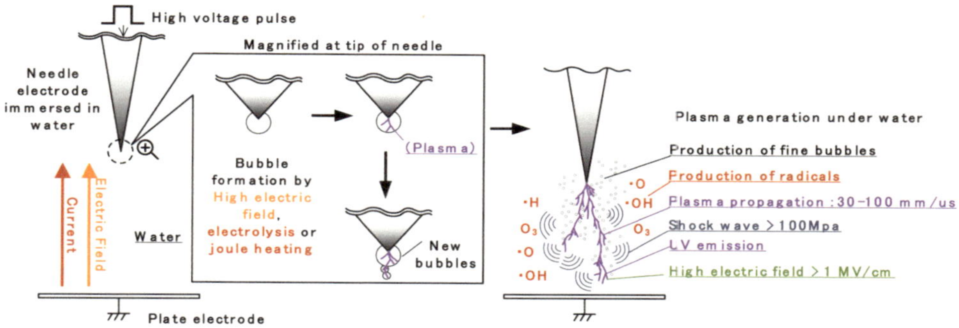

Figure 1.
Schematic illustration of discharge plasma using electrodes immersed in water.

conductivity due to the high concentration of various substances. When a voltage is applied to the electrodes immersed in water, which has high conductivity, a large current flow, and large energy are dissipated during Joule heating, which does not contribute to chemically active species production. The discharge volume decreases with increasing conductivity of the water at the same input energy. As a result, the energy efficiency for water treatment by discharge decreases to a low value under high-conductivity water conditions. Furthermore, because the breakdown strength of water is 300 MV/m and is 10 times higher than that of air (3.2 MV/m), high voltage is required to generate plasma discharges under water and to increase plasma volume. To solve this problem, methods to generate a pulsed discharge in contact with water using an electrode in the gas phase have been investigated [20–22]. Because the electrode is insulated by gas, the energy loss caused by the conductive current is significantly reduced. Plasma generation in the gas phase is much easier than that in the liquid phase, which contributes to a decrease in operating voltage and makes the system compact and light.

Various methods to generate a pulsed discharge in contact with water have been proposed [23, 24]. **Figure 2** shows schematic illustrations and photographs of

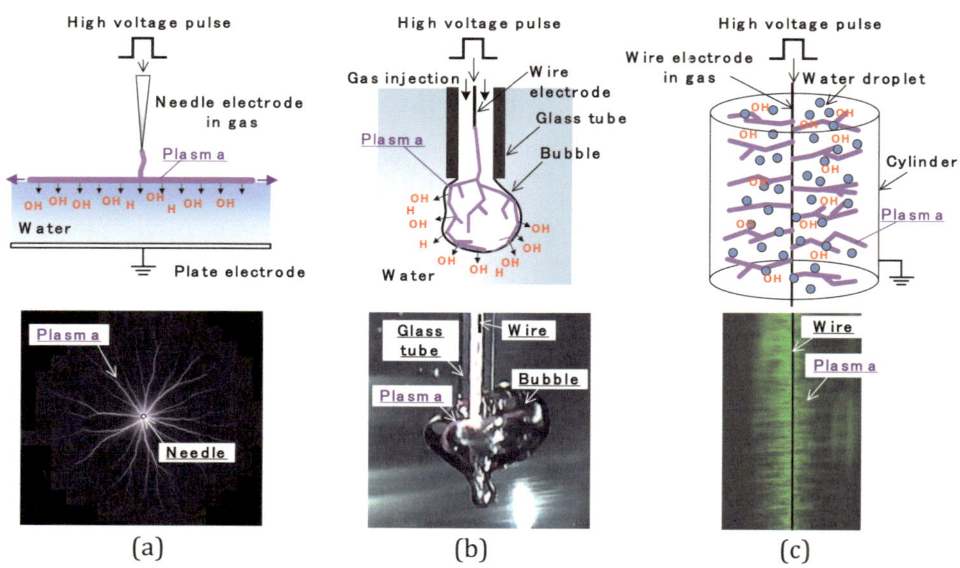

Figure 2.
Schematic illustrations and photographs of pulsed discharge in contact with water generated (a) above the stationary water surface, (b) inside the bubble, and (c) area where droplets are sprayed.

examples. The simplest method is to generate discharge that propagates and spreads over the water surface by using a high-voltage electrode placed above stationary water, [25, 26] as shown in **Figure 2(a)**. The chemical species generated in the plasma discharge are dissolved into the water and react with organic compounds in the water as described later. The discharges generated inside bubbles as shown in **Figure 2(b)**, which are produced by injecting gas into the water using a gas feed tube, have been widely investigated [23, 26–28]. This method has advantages such as a high ratio of water surface to gas volume, easy control of the gas purity and components, easy use in various fields, and automatic water circulation. The reactor can be simply constructed as shown in **Figure 3(a)** [29]. This reactor consists of a sealed glass vial for gas chromatography and glass tubes in which wire electrodes are inserted. Because the glass vial is sealed, the purity of the gas component in the reaction area in the vial can be improved. The treatment speed and volume can be increased easily by increasing the number of gas feed tubes and electrodes, as shown in **Figure 3(b)**. **Figure 3(c)** shows a reactor consisting of a separator film that has tiny pores and a high voltage wire electrode in the gas phase [20]. Multiple

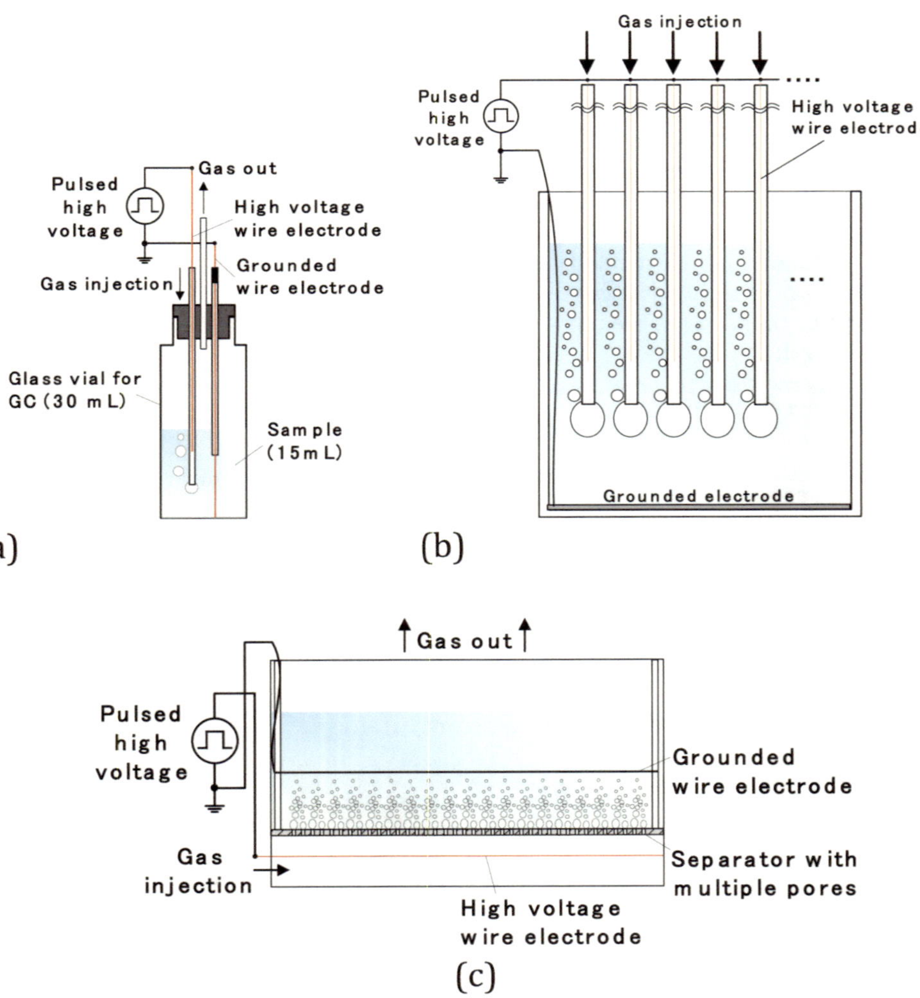

Figure 3.
A reactor used for pulsed discharge generation inside bubbles using (a) a single electrode inserted in a sealed glass vial, (b) parallel electrodes inserted in a vessel, and (c) a separator with multiple pores and a high voltage wire electrode placed in the gas phase.

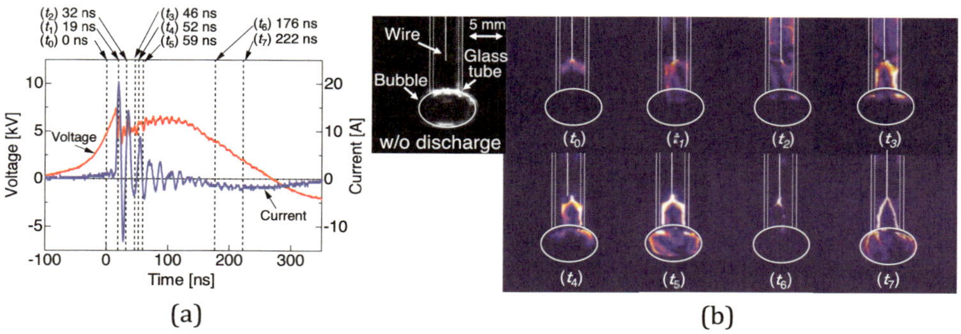

Figure 4.
(a) Typical voltage–current waveforms and (b) flaming photograph of discharge generated inside argon bubble.

discharge plasmas occur simultaneously on the wire electrode surface, which propagates into the bubbles. The separator film is used to separate the liquid phase and gas phase for insulation of the high voltage electrode. Instead of the separator and the wire electrode, ceramics with millipore needle electrodes have been used in some studies [7, 30].

When a pulsed high voltage is applied to the high-voltage electrode in the gas phase, the discharge propagates into the bubble surface from the high-voltage electrode [31, 32]. **Figure 4** shows typical voltage–current waveforms and flaming photographs of the pulsed discharge generated in an argon bubble, as shown in **Figure 2(b)** taken by using an intensified CCD camera with an exposure time of 5 ns. At time t_0, a corona discharge occurs at the tip of the electrode, and then, the discharge propagates to the tip of the glass tube along the surface of the glass at t_1 and t_2. Inside the glass tube, discharge occurs at not only the tip of the electrode but also the side of the electrode, as observed with dielectric discharges. At t_3, the discharge propagates along the bubble surface and does not travel to the bubble at t_3 to t_5. The results obtained by numerical simulation also show that a large part of the discharge inside the bubble propagates along with the bubble with similar size of bubbles [33, 34]. The propagation velocity of the discharge is 2.7–3.6 $\times$ 10^5 m/s in the Ar gas bubble [32] and 5 $\times$ 10^5 m/s on the surface of a glass tube filled with air, [31] which correspond to the surface discharge over a dielectric material under atmospheric pressure [35, 36]. When the voltage drops, the discharge disappears at t_6, but back discharge occurs again at t_7 due to the electric field formed by charges accumulated over the bubble surface.

Conductivity is the most important parameter for discharge propagation. Generally, the energy efficiency for discharge generation and propagation over a water surface is strongly affected by the conductivity of the water [32, 37]. **Figure 5** shows the equivalent circuit model [38] of (a) discharge in contact with water using the electrode system shown in **Figure 2(b)** discharge generated under water using the electrode system shown in **Figure 1**. When a breakdown occurs, switch SW_1 is turned on. The plasma impedance $Z_{P(t)}$ depends on the electron density of the plasma and changes with time during voltage application. The equivalent circuit of the water surface is expressed as the resistor R_L and capacitor C_L connected in parallel. The resistance of the resistor decreases with increasing conductivity. The discharge in contact with water propagates with the charging water surface. The decrease in resistance enhances the leakage of the accumulated charge on the surface of the bubble with decreasing conduction relaxation time, which is expressed by dividing the permittivity by the conductivity of the solution, which concentrates the discharge current and decreases the electric field. Therefore, the

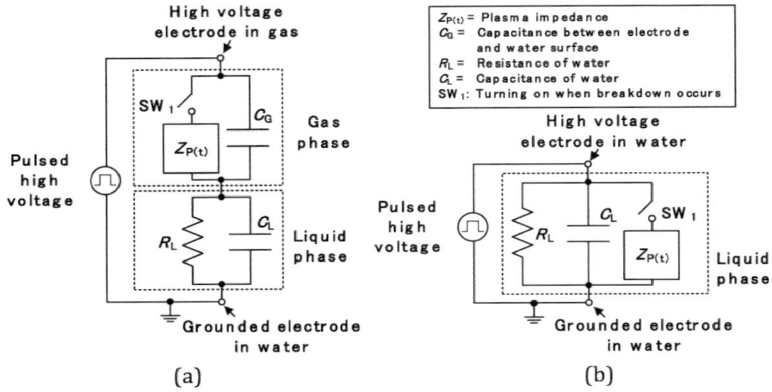

Figure 5.
Equivalent circuit model of (a) plasma contacted in water and (b) plasma generated under water.

discharge length decreases with increasing conductivity. For example, in the case of pulsed discharge generated inside a bubble as shown in **Figure 4**, the maximum discharge length decreases by 50% by increasing conductivity from 7 μS/cm to 1000 μS/cm [32].

When the electrode is placed in the gas phase and is not in contact with water, conductive current does not flow into the water without plasma generation, *i.e.*, without SW_1 turning on. On the other hand, when the electrode is placed in water, conductive current always flows into the water, which causes energy loss, and the energy efficiency for active species production decreases. This is one of the reasons that the energy efficiency of a wastewater treatment method using plasma contacted water is much higher than that using plasma directly generated under water [39]. Using this method, the energy efficiency can be improved over tens of times, especially for highly conductive wastewater treatment [39].

To reduce the influence of the conductivity, a method to generate streamer discharges into the area where droplets are sprayed using a pulsed high voltage is proposed as shown in **Figure 2(b)** [40]. This is one of the most efficient methods to supply chemical species into the water because streamer generation in the gas phase is not affected by water droplets, and the contact area between the discharge and water is large.

3. Production of chemical species by discharge plasma

In the plasma generated in the gas phase, various chemically active species such as ozone (O_3), oxygen radicals (O), and hydroxyl radicals (OH) are produced and dissolved at the water surface. Since these species have a high oxidation potential, they can contribute to wastewater treatment, *i.e.*, the decomposition of organic compounds and the inactivation of bacteria. The production reactions are initiated by the impacts of high-energy electrons on neutral molecules. Representative reactions in a gas that contains O_2, N_2, and Ar are as follows: [41–43].

$$O_2 + e\ (>8.4\ \text{eV}) \rightarrow O(^1D) + O(^3P) + e \quad (1)$$

$$O_2 + e\ (>5.58\ \text{eV}) \rightarrow O(^3P) + O(^3P) + e \quad (2)$$

$$O(^3P) + O_2 + M \rightarrow O_3 + M \quad (3)$$

$$H_2O + O(^1D) \rightarrow 2OH \quad (4)$$

$$N_2 + e\ (>9.0\ eV) \rightarrow N(^4S) + N(^2D) + e \quad (5)$$

$$N(^2D) + H_2O \rightarrow OH + NH \quad (6)$$

$$N(^2D) + O2 \rightarrow NO + O(1D) \quad (7)$$

$$N(^2D) + O_2 \rightarrow NO + O(^3P) \quad (8)$$

$$Ar + e\ (>11.55\ eV) \rightarrow Ar(^3P) + e \quad (9)$$

$$Ar(^3P) + H_2O \rightarrow Ar + OH + H \quad (10)$$

$$H_2O + e\ (>7.1\ eV) \rightarrow OH + H + e \quad (11)$$

$$H_2O + e\ (>12.62\ eV) \rightarrow H_2O^+ + 2e \quad (12)$$

$$H_2O^+ + H_2O \rightarrow H_3O^+ + OH \quad (13)$$

where (3) is a three-body recombination reaction and a neutral molecule such as O_2, N_2 or Ar is the third collision partner (M), which takes part in energy absorption, but does not react chemically. Ar, a noble gas, is generally used for highly efficient plasma generation at low cost. Because the reaction rates of (11) and (12) are not high in atmospheric nonthermal plasma, OH is mainly produced by indirect reactions such as reactions (4), (6) and (10) [44–46]. When a water surface is used as the cathode, OH is also produced under water by the bombardment of high-energy positive ions produced inside plasma, such as H_2O^+ and Ar^+, to the water surface [47, 48].

$$Ar + e\ (>15.8\ eV) \rightarrow Ar^+ + 2e \quad (14)$$

$$Ar^+ + H_2O \rightarrow H_2O^+ + Ar \quad (15)$$

$$H_2O^+ \rightarrow OH + H^+ \quad (16)$$

Figure 6 shows typical emission spectra of the discharge inside the Ar or O_2 bubble. The peaks at wavelengths of 282 and 308 nm are the $A^2\Sigma^+$ (v = 1) → $X^2\Pi$ (v = 0), and $A^2\Sigma^+$ (v = 0) → $X^2\Pi$ (v = 0) emission lines of OH, respectively. The peak at 656 nm is attributed to a Balmer H-alpha emission [32]. These peaks are strongly observed in argon gas because of its high electron density. In O_2 gas, the peak at a wavelength of 774 nm corresponds to the atomic oxygen (O) transition of

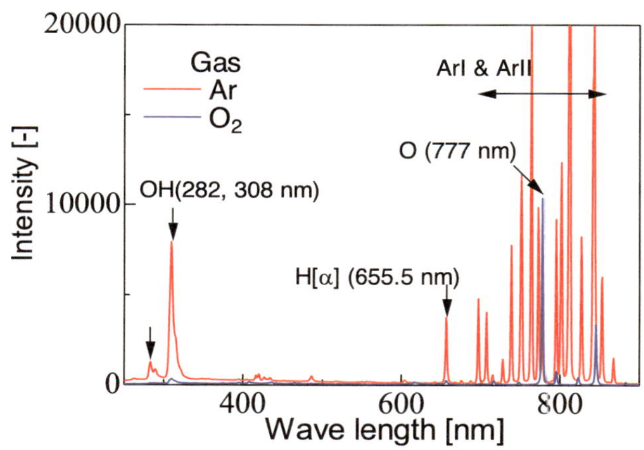

Figure 6.
Typical emission spectra of discharge inside argon or oxygen bubble.

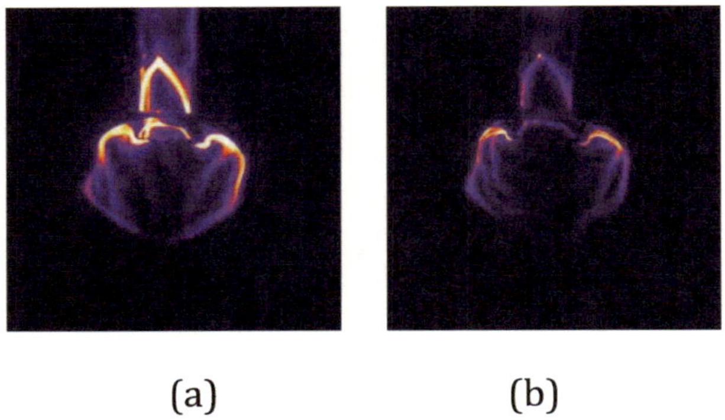

Figure 7.
Photograph of discharge inside the bubble (a) without and (b) with a bandpass filter (310 ± 10 nm).

$3p^5P$-$3s^5S^0$. **Figure 7** shows the photograph of discharge inside an Ar bubble taken by a high-speed intensified CCD camera with and without a bandpass filter with a bandwidth of 310 ± 10 nm, showing light emission with a wavelength corresponding to an excitation line of OH (308 nm). The photograph shows that the excitation-emission of OH with a wavelength of 308 nm is observed at the same position of the plasma channel generated along the bubble surface observed without the bandpass filter. This result shows that OH is produced inside the plasma channel.

Because these oxidants have high reactivities, they are consumed in loss reactions in both the gas and aqueous phases, which decreases the efficiency of wastewater treatment as shown in **Figure 8**. OH is well known as the most powerful oxidant and has a very short lifetime. OH is mainly lost by three-body recombination reactions, as shown in reaction (17) in the gas phase and reactions (18) and (19) in the aqueous phase [49, 50].

$$OH + OH + M \rightarrow H_2O_2 \tag{17}$$

$$OH + OH \rightarrow H_2O_2 \tag{18}$$

$$OH_) + H \rightarrow H_2O \tag{19}$$

The lifetime and diffusion constant of OH in the gas phase are on the order of 10^{-5} s and 10^{-10} m^2/s, respectively, and the diffusion length is several tens of μm [51]. The lifetime of dissolved OH is on the order of $10^{-6} \sim 10^{-7}$ s with penetration lengths on the order of $10^{-5} \sim 10^{-6}$ m [52–54]. **Figure 9** shows the hydrogen peroxide (H_2O_2) concentration produced by the OH recombination reaction, as

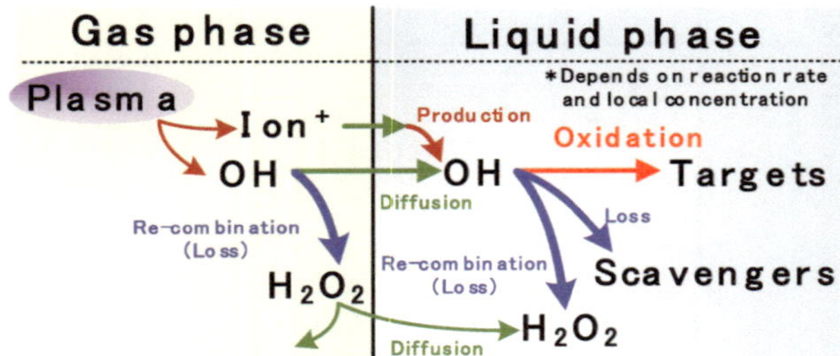

Figure 8.
Schematic illustrations of reactions of OH produced by plasma at the vicinity of water surface.

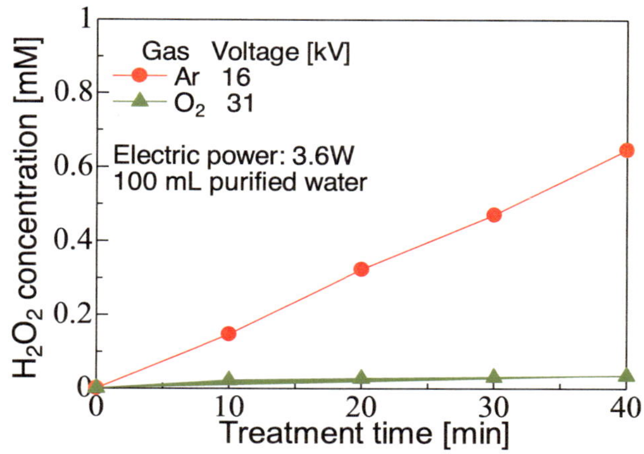

Figure 9.
Time change of the H_2O_2 concentration produced by plasma treatment.

described later, as a function of treatment time by plasma generated inside Ar or O_2 bubbles. In this case, 100 mL of dye solution was treated by plasma discharges generated inside bubbles produced by multiple pores, as shown in **Figure 3(c)**. The gas flow rate is 4.5 L/min. Pulse voltages with an amplitude of 20 kV, a pulse width of 160 ns, and a pulse repetition rate of 250 Hz are applied to the wire electrode placed in the gas phase to generate plasma [73]. The electric power into the plasma is the same in both gas species at 3.4 W. H_2O_2 concentration, *i.e.*, OH production, with Ar gas is higher than that with O_2 gas, which is in good agreement with the emission spectra, as shown in **Figure 6**. Thus, Ar injection is preferred for highly efficient OH production.

Many researchers have analyzed the amount of dissolved OH in water using chemical probes such as terephthalic acid [55, 56]. The energy yield for the surface pulsed streamer discharge on the liquid with argon gas is estimated to be $13 \sim 27 \times 10^{-9}$ mol/J with disodium terephthalate solution [57]. The typical energy yield of dissolved OH in water supplied by plasma, which can react with solutes in water, is estimated to be $4 \sim 5 \times 10^{-9}$ mol/J using the indigo carmine solution as a chemical probe [29]. Because plasma directly generated under water has a high OH density, the reaction rate of OH loss reactions is very high, which decreases the energy efficiency for OH supply.

When the feeding gas contains O_2, highly concentrated O_3 is generated in the plasma through reactions (1) ~ (3). Henry's law constants for O_3 and OH are 0.011 and 30 M/atm at 298 K, [58] respectively, which shows that the solubility of O_3 is lower than that of OH. Additionally, the oxidation potential of O_3 is lower than that of OH. However, the lifetime of O_3 in water is on the order of several min [59–61] and much longer than that of OH. Thus, when wastewater contains only organic compounds that can be easily oxidized by O_3, the treatment efficiency utilizing O_3 is higher than that utilizing OH. In this case, O_2 is preferred for use as a feeding gas for high-efficient O_3 production. O_3 dissolved in water can contribute to OH production via the following reaction with hydroperoxide anion (HO_2^-) species, the conjugate base of H_2O_2, under alkaline conditions with a high reaction rate [62].

$$O_3 + H_2O_2 \rightarrow O_2 + OH + HO_2 \quad (20)$$

$$H_2O_2 \leftrightarrow HO_2^- + H^+ \ pKa = 11.6 \quad (21)$$

$$O_3 + HO_2^- \rightarrow O_2 + OH + O_2^- \quad (22)$$

The pKa value of the equilibrium reaction (R14) is 11.6. Because the reaction rate of O_3 and H_2O_2 is very low (10^{-2} M^{-1} s^{-1}), O_3 mainly reacts with HO_2^- with a reaction rate constant of 5.5×10^6 M^{-1} s^{-1} and hardly reacts with H_2O_2 directly. Thus, pH adjustment is required to induce this OH production reaction.

When the feeding gas contains nitrogen, nitrous (HNO_2) and nitrite (HNO_3) acids, are generated by plasma as byproducts through the following reactions; [46, 63–65].

$$O(^3P) + NO + N_2 \rightarrow NO_2 + N_2 \tag{23}$$

$$O_3 + NO \rightarrow NO_2 + O_2 \tag{24}$$

$$NH + OH \rightarrow NO + H_2 \tag{25}$$

$$OH + NO + M \rightarrow HNO_2 \tag{26}$$

$$OH + NO_2 \rightarrow HNO_3 \tag{27}$$

$$3HNO_2 \rightarrow HNO_3 + 2NO + H_2O \tag{28}$$

$$HO_2 + NO_2 \rightarrow O_2NOOH \tag{29}$$

$$H_2O_2 + HNO_2 \rightarrow HOONO + H_2O \tag{30}$$

$$HOONO + H_2O_2 \rightarrow O_2NOOH + H_2O \tag{31}$$

Peroxynitric acid ($HOONO_2$) contributes to HOO production via equilibrium decomposition reaction, [66].

$$HOONO2 \rightarrow HOO + NO_2 \tag{32}$$

When HNO_2 and HNO_3 are dissolved in water, nitric and nitrate ions are easily produced in water through acid dissociations of HNO_2 and HNO_3 because of their low pKa values, of 3.35 and $-$ 1.8, respectively, and the pH of the solution significantly decreases [65, 67].

$$HNO_2 \rightarrow H^+ + NO_2^- \tag{33}$$

$$HNO_3 \rightarrow H^+ + NO_3^- \tag{34}$$

NO_2^- is oxidized to NO_3^- by O_3 in solution [68]. When the pH is lower than 3.5, NO_2^- is oxidized by H+. These reactions are follows:

$$NO_2^- + O_3 \rightarrow NO_3^- + O_2 \tag{35}$$

$$NO_2^- + 3H^+ \rightarrow NO + NO_3^- + H_3O \text{ (or } H_2O + H^+) \tag{36}$$

These nitrogen species can contribute to the inactivation of bacteria in wastewater as described later. In the case of plasma discharges inside multiple air bubbles using a reactor similar to that in **Figure 3(b)** with eight glass tubes in which total air gas with a flow rate of 5 L/min is injected, approximately 9 mM NO_3^- is generated in 20 L of solution with 24 hours of plasma treatment, [69] which is a sufficient concentration for using fertilizers in plant cultivation. This effect is described in Section 5.

When metals that are easily oxidized by electrolysis, such as iron and copper, are used as an electrode immersed in water, metal ions such as Fe^{2+}, Fe^{3+}, Cu^{+}, and Cu^{2+} are generated by electrooxidation by a large discharge current, and the metal ion concentration increases. These metal ions contribute to the oxidation of organic compounds in wastewater not only directly but also through OH production by redox reactions well known as Fenton reactions with H_2O_2 produced by plasma as follows: [70–73].

$$Fe^{2+} + H_2O_2 \rightarrow OH + OH^- + Fe^{3+} \quad (37)$$

$$Fe^{2+} + OH \rightarrow Fe^{3+} + OH^- \quad (38)$$

$$Fe^{3+} + H_2O_2 \rightarrow Fe^{2+} + HO_2 + H^+ \quad (39)$$

$$Fe^{3+} + HO_2 \rightarrow Fe^{2+} + O_2 + H^+ \quad (40)$$

$$Fe^{2+} + HO_2 \rightarrow Fe^{3+} + HO_2^- \quad (41)$$

$$Fe^{3+} + e^- \rightarrow Fe^{2+} \quad (42)$$

$$H + Fe^{3+} \rightarrow Fe^{2+} + H^+ \quad (43)$$

$$Cu^+ + OH \rightarrow Cu^{2+} + OH^- \quad (44)$$

$$O_2^- + O_2^- + 2H^+ \rightarrow H_2O_2 \quad (45)$$

Reactions related to iron ions are well known as Fenton reactions and occur under acidic conditions. Generally, electrodes immersed in water are used as grounded electrodes to keep the voltage potential of water zero for safety reasons. Therefore, when a positive (or negative) high voltage is applied to a high voltage electrode, iron (or copper) ions are dissolved from the grounded electrode immersed in water and used as the anode (or cathode). For example, in the case of 1,4-decomposition as described in Section 4, the decomposition efficiency can be improved by approximately three times, by using iron wire as an electrode immersed in water, compared with that achieved using stainless steel wire [73].

The chemical species dissolved in the water react with not only organic compounds but also inorganic compounds. **Table 1** shows reaction the rate constants for

Solute	Reaction rate constant for O_3 [$M^{-1}s^{-1}$]	Reaction rate constant for OH [$M^{-1}s^{-1}$]
Cl^-	0.003	4.3×10^9
Br^-	160	1.1×10^9
I^-	1.2×10^5	1.1×10^{10}
NO_2^-	3.7×10^5	1.0×10^{10}
NO_3^-	—	1.0×10^5
HSO_3^-	3.7×10^5	4.5×10^9
SO_3^{2-}	1.5×10^9	5.1×10^9
HSO_4^-	$<10^{-4}$	6.9×10^5
H_2SO_4	—	1.4×10^7
H_2O_2	0.01	2.7×10^7
HO_2^-	5.5×10^6	7.5×10^9
HCO_3^-	0.01	8.5×10^6
HCO_2^-	100	5.1×10^9
CO_3^{2-}	0.01	3.9×10^8
CO_2	—	1×10^6
OH^-	210	1.2×10^{10}

Table 1.
Reaction rate constant for O_3 and OH reacting with inorganic compounds.

O_3 and OH reacting with inorganic ions [74–78]. Halogen ions such as chloride ions (Cl^-), bromide ions (Br^-) and iodide ions (I^-) have high reactivity toward O_3 and OH and act as O_3 and OH scavengers, which inhibits the decomposition reactions of organic compounds. Furthermore, toxic compounds such as hypochlorite ions (ClO^-) and bromate ions (BrO_3^-) are produced via oxidation reactions, as shown in **Figures 10** and **11**. Carbon dioxide radicals (CO_2^-), carbonate radicals (CO_3^-), and

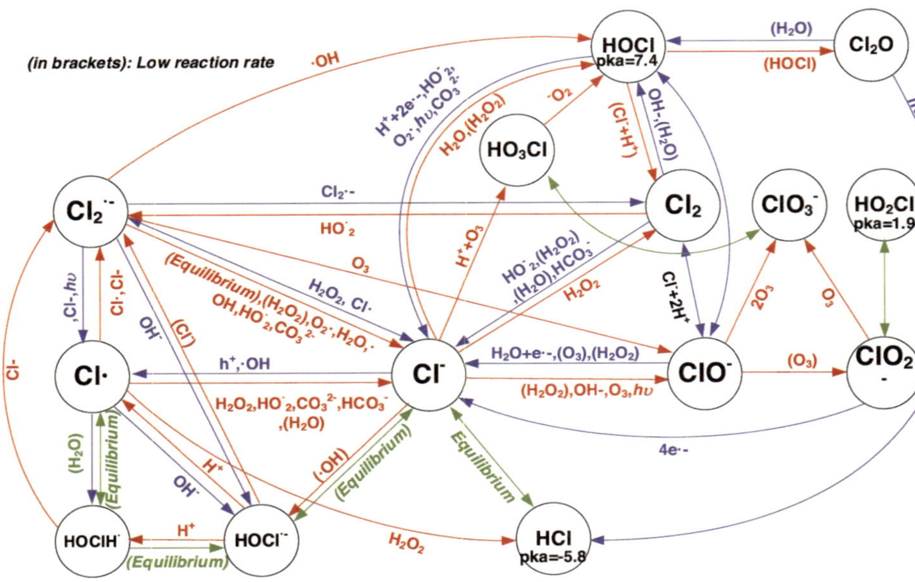

Figure 10.
Schematic illustrations of reactions involving Cl.

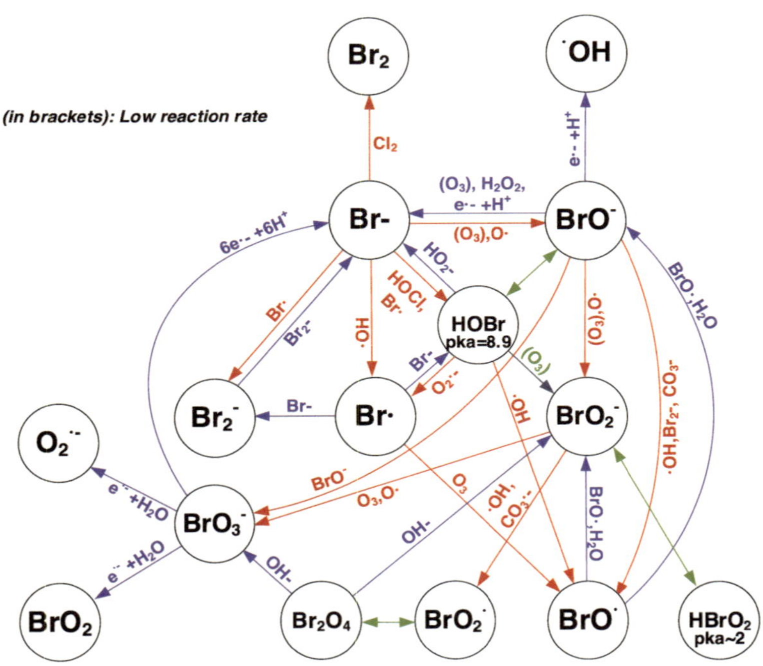

Figure 11.
Schematic illustrations of reactions involving Br.

sulfate radicals (SO_4^-) are produced through reactions of OH with formate ions (HCO_2^-), carbonate ions (CO_3^{2-}), and hydrogen sulfate ions (HSO_4^-) with hydroxyl radicals in liquid [79]. These radicals also contribute to the decomposition of organic compounds dissolved in wastewater as synergistic effects. Although hydroxide ions (OH^-) and per hydroxyl ions (HO_2^-) act as scavengers of OH, OH can be produced via reaction with O_3 under high-pH conditions. Thus, the influence of these ions on treatment efficiency and water quality after wastewater treatment using plasma should be considered.

4. Decomposition of persistent organic pollutants

Organic pollutants entering the water environment cause industrial, agricultural, domestic, livestock, and medical wastewater drainage into and leakage from polluted soil. Because most organic pollutants have strong toxicity, carcinogenicity, and genotoxicity, even at low concentrations, it is urgent to develop effective technologies for the highly efficient removal of these contaminants from water. 1,4-Dioxane has been widely used as a solvent. Because it has high solubility and is hard to absorb to the bottom material, it can be widely spread in water environments [80, 81]. Dichloromethane (DCM) is a chlorinated volatile organic compound (Cl-VOC) and has high volatility. It is also used as a solvent, cleaner, degreasing agent, and liquid fuel and released into the environment through water discharge from chemical industries [82, 83]. These compounds are very difficult to decompose using conventional oxidation treatment methods such as chlorine, ozone oxidation, and bioremediation owing to their chemical and biological stability.

In wastewater treatment systems using plasma discharges, high-density powerful oxidants such as OH and O_3 produced by plasma contribute to the decomposition of persistent organic pollutants and the inactivation of bacteria. This method provides high-speed treatment and decomposition of organic compounds that are difficult to remove with conventional methods and has attracted much attention as an advanced oxidation technology. Many researchers have proposed that plasma treatment is effective for the decomposition of various kinds of toxic aromatic hydrocarbons, such as chlorophenol, [84] naphthalene, toluene [85], and bisphenol A [86], and volatile organic compounds such as trichloroacetic acid [87]. Oxidants also contribute to the quick decolorization of wastewater containing dye solutions [88, 89]. In this section, the decomposition of various kinds of organic compounds in water is systematically described. The decomposition characteristics of indigo carmine, a common blue dye, 1,4-dioxane [73] which has a high solubility in water, and formic acid, [90] which is an intermediate product of various organic compounds, such as dichloromethane (DCM), [91] which is a volatile organic compound, are introduced as an example.

Chemically active species dissolve in water in the vicinity of the water surface and induce oxidation reactions such as hydrogen atom abstraction, electron transfer, electrophilic addition, and chain oxidation reactions triggered by these reactions. The following reactions are typical oxidation reactions of organic compounds induced by OH.

$$RH + OH \rightarrow R^{\bullet} + H_2O \quad (46)$$

$$R + OH \rightarrow R^{\bullet +} + OH^- \quad (47)$$

$$R - +OH \rightarrow R^{\bullet} + OH^- \quad (48)$$

$$R + OH \rightarrow {}^{\bullet}ROH \quad (49)$$

$$OH + R_2C = CR_2 \rightarrow R_2(OH)C - CR_2^{\bullet} \quad (50)$$

$$R^{\bullet} + O_2 \rightarrow RO_2^{\bullet} \quad (51)$$

$$R^{\bullet} + R^{\bullet} \rightarrow RR \quad (52)$$

$$R^{\bullet} + RO_2^{\bullet} \rightarrow RO_2R \quad (53)$$

$$RO_2^{\bullet} + RH \rightarrow ROOH + R^{\bullet} \quad (54)$$

$$RO^{\bullet} + RH \rightarrow ROH + R^{\bullet} \quad (55)$$

$$RO_2^{\bullet} + HO_2^{\bullet} \rightarrow RO^{\bullet} + O_2 + OH^{\bullet} \quad (56)$$

$$RO_2^{\bullet} + RO_2^{\bullet} \rightarrow RO2R \quad (57)$$

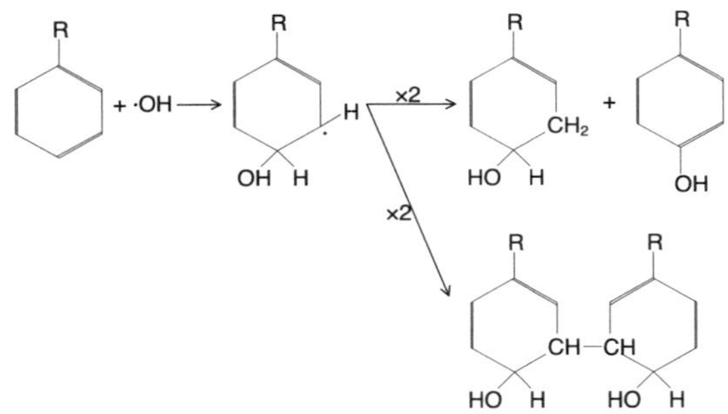

(58)

Indigo carmine and azo phloxine are common dyes and have blue and red colors, respectively. **Figure 12** shows the time change of appearance and UV–vis spectra of two dye solutions, (a) indigo carmine and (b) azo phloxine, by treatment using plasma generated inside O_2 gas bubbles. In this case, 100 mL of dye solution was treated by plasma discharges generated inside bubbles produced by multiple pores as shown in **Figure 3(c)**. The gas flow rate is 4.5 L/min. Pulse voltages with an amplitude of 20 kV, a pulse width of 160 ns, and a pulse repetition rate of 250 Hz are applied to the wire electrode placed in the gas phase to generate plasma [73]. The energy input into a pulse is fixed at 13.5 mJ for each gas species by adjusting the voltage amplitude. From the input energy and pulse repetition rate, the input power can be estimated as 3.4 W. Indigo carmine and azo phloxine have absorbance peaks at wavelengths of 610 nm and 531 nm, which are responsible for the blue and red colors of their solutions, respectively. These peaks decrease, the colors disappear after 20 s and 180 s of plasma treatment, and then the solutions are fully decolorized after treatment. This result shows that plasma treatment is effective for the decolorization treatment of wastewater. **Figure 13** shows the decolorization rate of these solutions using Ar and O_2 as feeding gases to generate bubbles. Decolorization of both solutions in the case of O_2 injection is much faster than that of Ar injection. The energy efficiency for 50% decolorization of indigo carmine is 1.3×10^3 mmol/kWh for O_2 injection and 6.1 mmol/kWh for Ar injection, and those of azo phloxine are 77 mmol/kWh for O_2 injection and 4.8 mmol/kWh for Ar injection. The absorbance peaks of these dyes are due to the carbon–carbon double bond in the H-chromophore of indigo carmine and the nitrogen-nitrogen double bond of the azo chromophore of azo phloxine as shown in **Figure 14**, which are responsible for the blue and red colors of the solution. These double bonds have high electron density and high reactivity to electrophilic OH. It is well known that O_3 reacts with C-C and N-N double bonds rapidly; however, the reaction of OH with dyes is at least four orders of magnitude faster than that of O_3. As already mentioned, the lifetime of OH in water is much shorter than that of O_3 because of its high reactivity and high rate constant of recombination loss reactions, and only OH generated in the vicinity

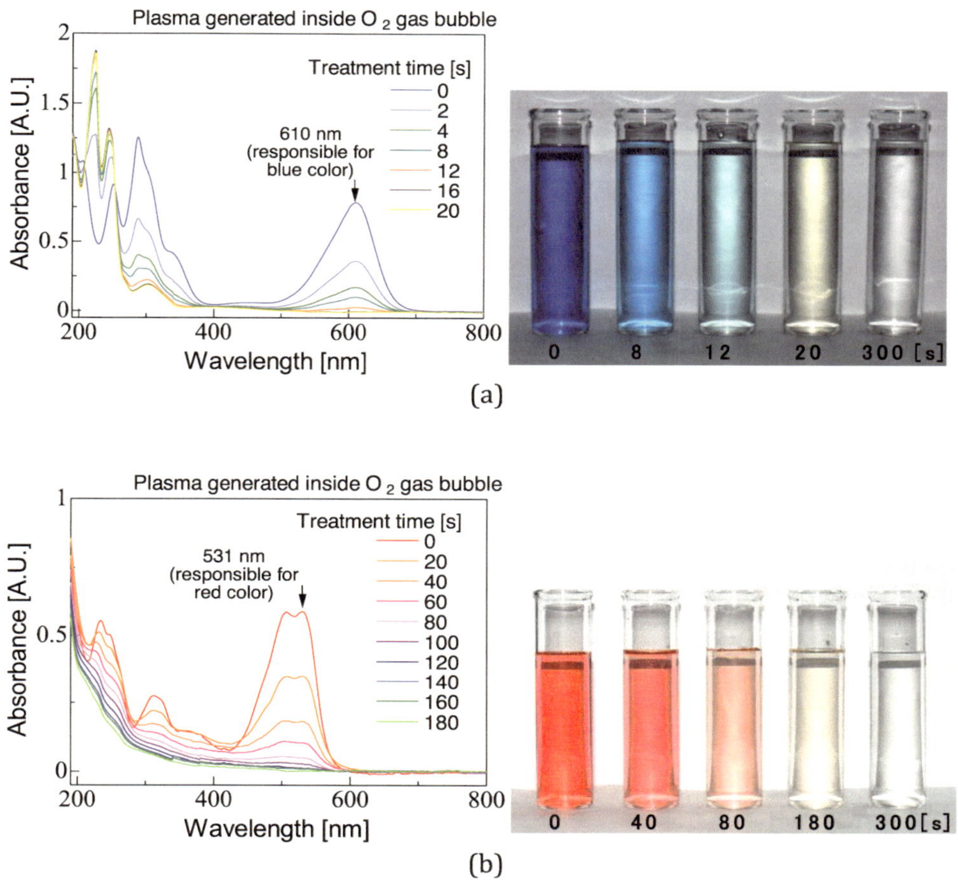

Figure 12.
Time change of appearance and UV–vis spectra of two dye solutions, (a) indigo carmine and (b) azo phloxine, by treatment using plasma generated inside O_2 gas bubble.

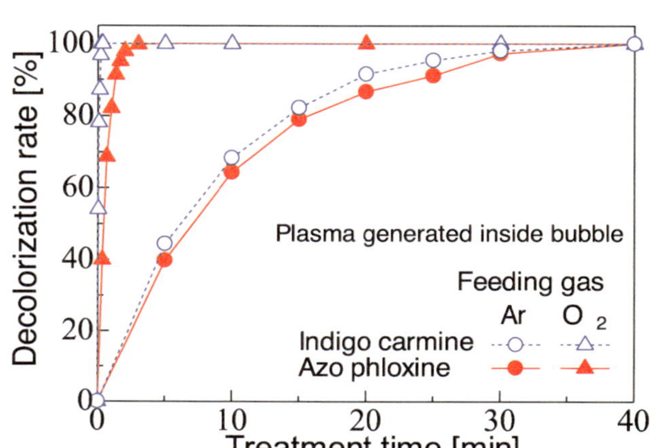

Figure 13.
Time change of decolorization rate of indigo carmine and azo phloxine solutions by plasma treatment using Ar and O_2 gases.

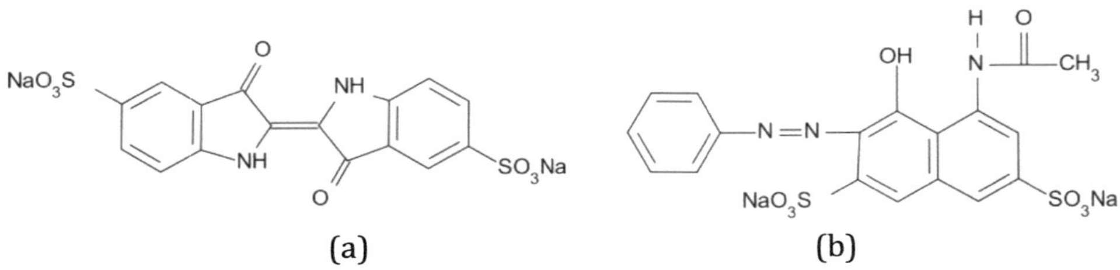

Figure 14.
Chemical structure of (a) indigo carmine and (b) azo phloxine.

of the bubble surface can be dissolved in the solution. In this case, these two dyes can be easily oxidized by O_3. Therefore, O_2 injection to generate O_3 at high efficiency is preferred for the decomposition of compounds that have high reactivity to O_3. The reactivity of an azo group with ozone is very low compared to that of olefins, [92] which is one of the reasons that indigo carmine is more readily decolorized than azo phroxine.

Since 1,4-dioxane has low reactivity to O_3 due to its high chemical stability, plasma treatment is a promising method for 1,4-dioxane removal. **Figure 15** shows one of the proposed decomposition processes of 1,4-dioxane by hydroxyl radicals [93–95]. The reaction is initiated by hydrogen atom abstraction, and various reactions, such as chain reactions, are involved. Carboxylic acids such as formic acids (HCOOH) and oxalic acid ($(COOH)_2$) are produced as intermediate products.

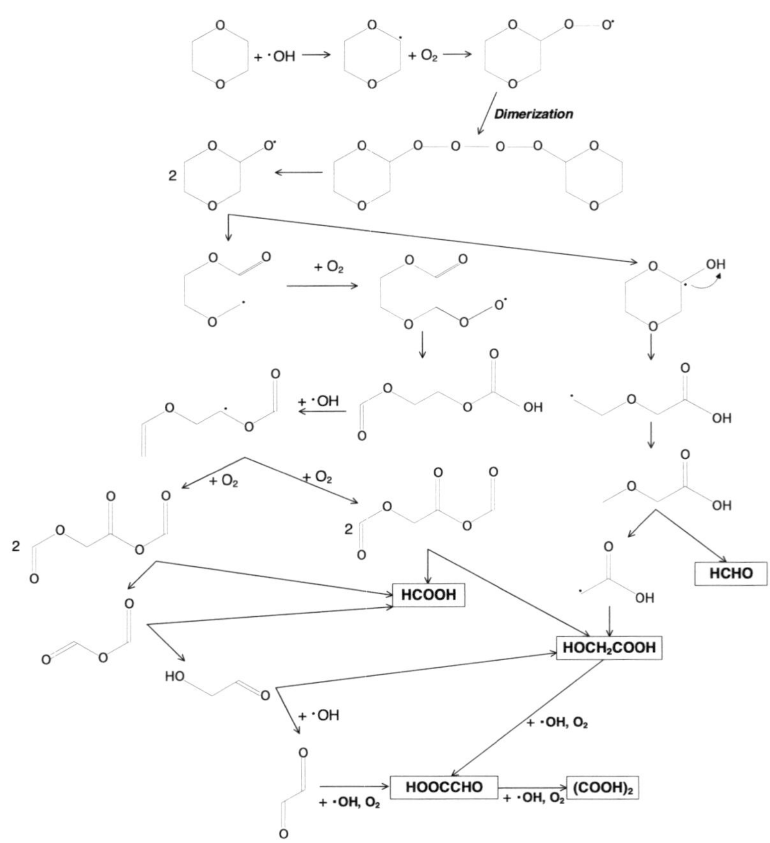

Figure 15.
Oxidation process of 1,4-dioxane by OH.

DOI: http://dx.doi.org/10.5772/intechopen.101494

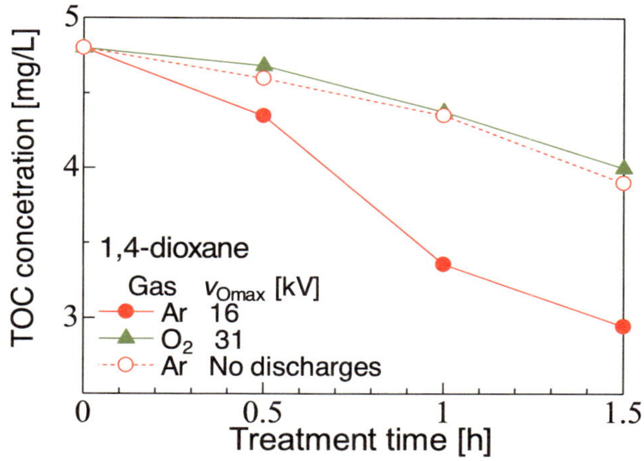

Figure 16.
Time change of the TOC concentration of a 1,4-dioxane solution by plasma treatment.

Figure 16 shows the time change of the TOC concentration of the 1,4-dioxane solution with an initial concentration of 0.23 mM with two different injected gases and with and without discharges. The experimental conditions were the same as those mentioned above. The electric power in the plasma is the same in both gas species. The TOC decreases without discharges because of the volatilization by bubbling gas, and the TOC concentration in the O_2 injection is almost the same without discharges. The rate of TOC decrease rate in the case of Ar injection is much higher than that in the case of O_2 injection. 1,4-dioxane hardly reacts with O_3 because of its low reaction rate (0.32 M^{-1} s^{-1}) [78]. The rate constant of reaction between OH and 1,4-dioxane is 2.4*10^9 M^{-1} s^{-1} [96]. The production rate of OH in the case of Ar injection is much higher than that in the case of O_2 injection. Thus, Ar injection is preferred for the decomposition of organic compounds that have

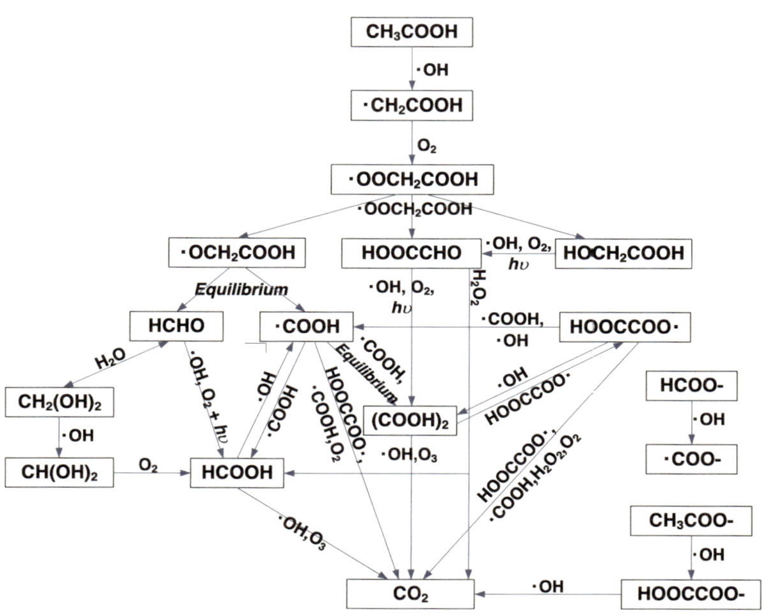

Figure 17.
Oxidation process of CH_3COOH by OH.

low reactivity to O_3. The energy efficiency for 1,4-dioxane decomposition is 2 mmol/kWh.

Figure 17 shows the oxidation of acetic acid (CH_3COOH) by OH radicals. CH_3COOH has also low reactivity to O_3. It can be decomposed into formic acid, oxalic acid, and finally, CO_2 and H_2O by oxidation reactions with OH. These acids have an acidity constant and are in equilibrium with their conjugated bases. For example, formic acid is in equilibrium with its conjugate base, formate, as follows:

$$HCOOH \rightleftarrows HCOO - +H+ \tag{59}$$

The pKa value of this equilibrium reaction is 3.75. The reaction processes of formic acid and formate with OH and O_3 are represented by the following: [76, 78].

$$HCOOH + O_3 \rightarrow CO_2 + OH + HO_2 \tag{60}$$

$$HCOOH + \cdot OH \rightarrow H_2O + COOH \tag{61}$$

$$HCOO^- + O_3 \rightarrow CO_2 + OH + O_2{}^- \tag{62}$$

$$HCOO^- + OH \rightarrow H_2O + COO^- \tag{63}$$

The rate constant of reaction (62) is 100 $M^{-1}\,s^{-1}$ and is 20 times higher than that of reaction (60) (5 $M^{-1}\,s^{-1}$). Similarly, the rate constant of reaction (63) is $3.2 \times 10^9\ M^{-1}\,s^{-1}$ and is 25 times higher than that of reaction (61) ($1.3 \times 10^8\ M^{-1}\,s^{-1}$). Therefore, decomposition of these acids at higher pH is preferred. It has been reported that the decomposition rate of formic acid can be improved twice by increasing the pH from 3.5 to 4.2 with the addition of sodium carbonate, in the case of plasma generated inside oxygen bubbles using the reactor shown in **Figure 3(a)** [90]. Thus, pH adjustment is an important factor for waste-water treatment, while active species can be directly supplied from plasma.

The decomposition process of DCM is different from that of other soluble compounds because of its high volatility. The rate constant of DCM reacting with OH in a liquid is $2.2 \times 10^7\ M^{-1}S^{-1}$ [97] and is much lower than that of reactions of other soluble organic compounds, such as 1,4-dioxane, phenol, and formic acids, with OH (on the order of $10^9\ M^{-1}\,s^{-1}$). **Figure 18** shows the TOC and the amount of decomposed DCM as a function of treatment time using plasma generated inside Ar

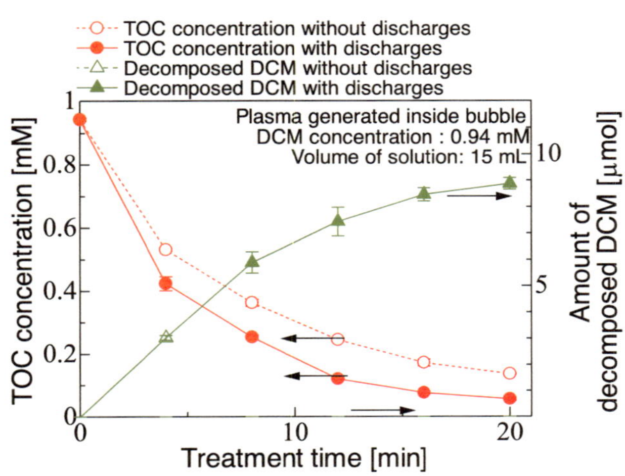

Figure 18.
Time change of the TOC concentration of DCM solution and amount of decomposed DCM by plasma treatment.

bubbles using the reaction shown in **Figure 3(a)**. The gas flow rate is 30 mL/min. Pulse voltages with an amplitude of 16 kV, a pulse width of 160 ns, and a pulse repetition rate of 250 Hz are applied to the wire electrode placed in the gas phase to generate plasma [91]. The amount of decomposed DCM is calculated from the Cl^- concentration detached from DCM during the decomposition process. The TOC decreases without discharges because of volatilization by bubbling gas. DCM can be successfully decomposed by discharges generated inside Ar bubbles [91]. The energy efficiency for DCM decomposition is approximately 60 mmol/kWh and is much higher than that for decolorization of dye solutions and 1,4-dioxane decomposition, as mentioned above, although the rate constant for DCM reacting with OH is low. This result suggests that almost all DCM is decomposed in the gas phase in the bubble, not in the liquid phase. DCM is quickly volatilized due to its high volatility and is transferred into the gas phase in the bubble when Ar gas is fed into the DCM solution, and DCM is exposed by plasma discharges generated inside the bubble. DCM can be decomposed by OH with the following reactions: [98].

$$CH_2Cl_{2(g)} + OH_{(g)} \rightarrow CHCl_{2(g)} + H_2O_{(g)} \tag{64}$$

$$CHCl_{2(g)} + OH_{(g)} \rightarrow CHOCl_{(g)} + HCl_{(g)} \tag{65}$$

$$CHOCl_{(g)} + OH_{(g)} \rightarrow COCl_{(g)} + H_2O_{(g)} \tag{66}$$

$$COCl_{(g)} + OH_{(g)} \rightarrow CO_{2(g)} + HCl_{(g)} \tag{67}$$

The Cl-H bonding energy of CH_2Cl_2 is 4.28 eV and is lower than the energy of metastable Ar (11.55 eV) produced by electron impact, as shown in reaction (9) [99]. Therefore, CH_2Cl_2 is also possibly decomposed by Penning ionization by the energy transfer collision of metastable Ar (Ar*) as shown in the following reactions: [99, 100].

$$CH_2Cl_{2(g)} + Ar*_{(g)} \rightarrow CH_2Cl_2{}^{+}{}_{(g)} + Ar_{(g)} \tag{68}$$

$$CH_2Cl_2{}^{+}{}_{(g)} \rightarrow CH_2Cl^{+}{}_{(g)} + Cl_{(g)} \tag{69}$$

$$CH_2Cl^{+}{}_{(g)} + e^{-}{}_{(g)} \rightarrow CH_2Cl_{(g)} + Cl_{(g)} \tag{70}$$

Because the concentration of DCM inside the gas bubble is on the order of several hundred ppm as mentioned previously, and is much lower than the Ar and H_2O concentrations, the DCM dissociation by electron impact is negligible [98, 101].

5. Novel applications for agriculture

Agricultural applications of plasma have been widely investigated in recent years and have become one of the most attractive research topics in plasma science [102–105]. As already mentioned in this article, plasma in contact with water can produce various types of chemically active species, which have various effects on the environment of plant growth.

As a production process of fruits and vegetables, hydroponics, which is the method of growing plants without soil using the nutrient solution, has been widely used. In hydroponics, the nutrient solution is recirculated in a closed system to reduce the cost and the environmental load. In the system, plant diseases caused by microbial contamination of artificial nutrient solution rapidly spread in the circulation system and cause serious damage to the entire plant. During the entire period of plant growth, contamination with pathogens can never be excluded since pathogens are introduced in the nutrient solution via the irrigation water supply. Therefore, the nutrient solution should be remedied by continuous water treatment system operation during the cultivation period. Active species such as OH and O_3 produced by discharges can contribute to the inactivation of pathogenic bacteria-contaminated in liquid fertilizer. A plasma treatment system for the inactivation of *Ralstonia solanacearum* bacteria in liquid fertilizer for a tomato hydroponic culture system has been developed [106]. The number of colony-forming units (CFU) of *R. solanacearum* in the liquid fertilizer decreased from 10^7 to 10^2 CFU/mL when treated with discharge plasma. Tomato seedlings treated with discharge plasma were relatively healthy, while the infected positive controls all wilted and died, as shown in **Figure 19**. Plasma was also used for the removal of allelochemicals, organic compounds exuded from the root of plants, which have autotoxic effects on plant growth. 2,4-Dichlorobenzoic acid (DCBA), an allelochemical of cucumber, was almost completely decomposed by plasma treatment. The discharge-treated solution is used as a nutrient solution for cultivating cucumber plants in a rockwool hydroponic system. Plant growth was significantly inhibited by adding DCBA to the nutrient solution, on the other hand, the plants grew healthy using discharge treatment, as shown in **Figure 20** [69]. Furthermore, when the feeding gas contains nitrogen, NO_2^- and NO_3^- are generated as mentioned already. These ions can act as

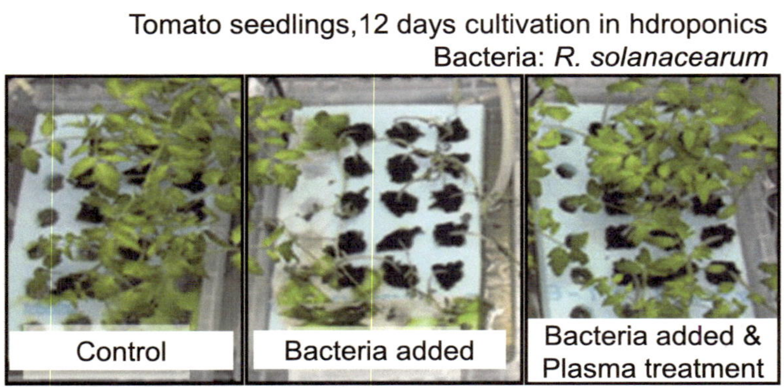

Figure 19.
Photograph of tomato seedlings after 12 days of cultivation in bacteria-contaminated nutrient medium in a hydroponic system.

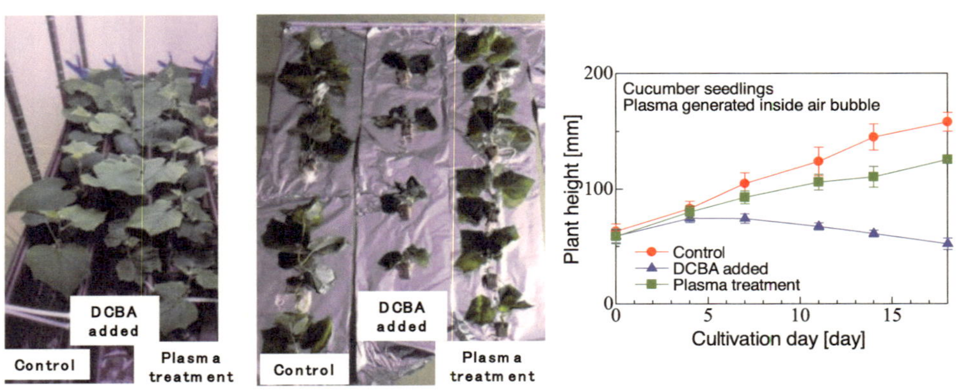

Figure 20.
Photographs of cucumber seedlings after 18 days and diurnal changes in plant height for seedling cultivated in DCBA-contaminated nutrient medium in a hydroponics system.

Figure 21.
Photograph of Japanese mustard spinach seedlings after 28 days of cultivation using plasma-treated nutrient medium-supplied soil.

fertilizers for plants, which improves the growth rate of plants in hydroponic systems, as shown in **Figure 21** [107]. Therefore, plasma treatment of nutrient solution for plant cultivation can control the environment of plant growth through the decomposition of toxic organic compounds, the inactivation of pathogenic bacteria, and the nutrient supply, which are promising new applications of plasma. A plasma treatment system for tomato cultivation in a greenhouse was developed, and a pilot test was performed over several months [108]. The CFU of *R. solanacearum* was kept at a low value, and the infection risk of the plant was significantly reduced during continuous treatment. All plant bodies grow healthily and bear fruits, and the onset of bacterial wilt disease is not observed.

Not only radicals such as O_3 and OH, but also nitrogen species such as $ONOO^-$ and O_2NOOH can be key inactivation agents that cause cell damage and inactivation under acidic conditions. Because these nitrogen species have a long lifetime in the liquid phase, nitrogen species generated by plasma treatment remain for a long time after nonthermal plasma irradiation. It has been reported that nonthermal plasma-activated water (PAW), water irradiated by nonthermal plasma, can inactivate bacteria on the surface of fruits and vegetables [109] such as strawberries, [110] mushrooms [111] and endives, [112] which contributes to maintaining the freshness of fruits and vegetables.

6. Conclusion

The fundamental physical and chemical properties of plasma discharge generated over a water surface and its applications for the decomposition of persistent compounds are described. Plasma can directly generate powerful oxidizing agents such as hydroxyl radicals, which can realize high-speed decomposition of persistent compounds in wastewater, such as 1,4-dioxane and dichloromethane, without pretreatment of samples. Because of these advantages over other advanced oxidation processes, a pulsed discharge plasma in water and in contact with water has attracted much attention as a promising technology in not only wastewater treatment but also various new application fields such as agriculture.

Authors are grateful to Prof. Y. Minanitani in Yamagata University, Japan for providing a photograph of discharge used in **Figure 2(c)**. This research was supported by a grant-in-aid from the Ministry of Agriculture, Forestry and Fisheries of Japan (A scheme to revitalize agriculture and fisheries in disaster areas through deploying highly advanced technology), by a grants-in-aid for Scientific Research (S), (A), (B), (C) and early-career scientists, from the Japan Society for the Promotion of Science, Grant Numbers 19H05611, 15H02231, 20,370,130, 21 K03994, 18 K13735 and respectively, and by Joint Usage/Research by the Institute of Pulsed Power Science, Kumamoto University.

Conflict of interest

The authors declare no conflict of interest.

Author details

Katsuyuki Takahashi[1,2]*, Koichi Takaki[1,2] and Naoya Satta[3]

1 Faculty of Science and Engineering, Iwate University, Morioka, Japan

2 Agri-Innovation Center, Iwate University, Morioka, Japan

3 Department of Architecture, Daido University, Nagoya, Aichi, Japan

*Address all correspondence to: ktaka@iwate-u.ac.jp

References

[1] Attri P, Sarinont T, Kim M, Amano T, Koga K, Cho AE, et al. Influence of ionic liquid and ionic salt on protein against the reactive species generated using dielectric barrier discharge plasma. Scientific Reports. 2015;**5**:17781. DOI: 10.1038/srep17781

[2] Nakajima A, Uchida G, Kawasaki T, Koga K, Sarinont T, Amano T, et al. Effects of gas flow on oxidation reaction in liquid induced by He/O2 plasma-jet irradiation. Journal of Applied Physics. 2015;**118**:043301. DOI: 10.1063/1.4927217

[3] Graves DB. The emerging role of reactive oxygen and nitrogen species in redox biology and some implications for plasma applications to medicine and biology. Journal of Physics D: Applied Physics. 2012;**45**:263001. DOI: 10.1088/0022-3727/45/26/263001

[4] Huang Y, Kou Y, Zheng C, Xu Y, Liu Z, Yan K. Escherichia coli inactivation in water using pulsed discharge. IEEE Transactions on Plasma Science. 2016;**44**:2654-2655. DOI: 10.1109/TPS.2016.2559802

[5] Kirkpatrick MJ, Dodet B, Odic E. Atmospheric pressure humid argon DBD plasma for the application of sterilization- measurement and simulation of hydrogen, oxygen, and hydrogen peroxide formation. International Journal of Plasma Environmental Science and Technology. 2007;**1**:96-101

[6] Miklos DB, Remy C, Jekel M, Linden KG, Hübner U. Evaluation of advanced oxidation processes for water and wastewater treatment – A critical review. Water Research. 2018;**139**:118-131. DOI: 10.1016/j.watres.2018.03.042

[7] Takeuchi N, Yasuoka K. Review of plasma-based water treatment technologies for the decomposition of persistent organic compounds. Japanese Journal of Applied Physics. 2021;**60**:SA0801. DOI: 10.35848/1347-4065/abb75d

[8] Qin L, Kim K, Takahashi K, Kang J, Takeuchi N, Li OL. N2/Ar plasma-induced surface sulfonation on graphene nanoplatelets for catalytic hydrolysis of cellulose to glucose. Applied Surface Science. 2021;**545**:149051. DOI: 10.1016/j.apsusc.2021.149051

[9] Takahashi K, Saito Y, Oikawa R, Okumura T, Takaki K, Fujio T. Development of automatically controlled corona plasma system for inactivation of pathogen in hydroponic cultivation medium of tomato. Journal of Electrostatics. 2018;**91**:61-69. DOI: 10.1016/j.elstat.2017.12.006

[10] Takaki K, Hayashi H, Wang D, Ohshima T. High-voltage technologies for agriculture and food processing. Journal of Physics D: Applied Physics. 2019;**52**:473001. DOI: 10.1088/1361-6463/ab2e2d

[11] Tanaka H, Ishikawa K. State of the art in medical applications using non-thermal atmospheric pressure plasma. Reviews of Modern Plasma Physics. 2017;**1**:1-89. DOI: 10.1007/s41614-017-0004-3

[12] Fridman A, Chrokov A, Gutsol A. Non-thermal atmospheric pressure. Journal of Physics D: Applied Physics. 2005;**38**:R1-R24. DOI: 10.1088/0022-3727/38/2/R01

[13] Huiskamp T. Nanosecond pulsed streamer discharges Part I : Generation , source-plasma interaction and energy-ef fi ciency optimization. Plasma Sources Science and Technology. 2020;**29**:023002. DOI: 10.1088/1361-6595/ab53c5

[14] Wang D, Namihira T. Nanosecond pulsed streamer discharges : II . Physics , discharge characterization and plasma processing. Plasma Sources Science and Technology. 2020;**29**:023001. DOI: 10.1088/1361-6595/ab5bf6

[15] Mankowski J, Kristiansen M, Fellow L. A review of short pulse generator technology. IEEE Transactions on Plasma Science. 2000; **28**:102-108. DOI: 10.1109/27.842875

[16] Wetz DA, Mankowski JJ, Dickens JC, KristiansenM. The impact of field enhancements and charge injetion on the puled breakdown strength of water. IEEE Transactions on Plasma Science. 2006;**34**:1670-1679. DOI: 10.1109/TPS.2006.881891

[17] Gavrilov IM, Kukhta VR, Lopatin VV, Petrov PG. Dynamics of prebreakdown phenomena in a uniform field. IEEE Transactions on Dielectrics and Electrical Insulation. 1994;**1**: 496-502. DOI: 10.1109/94.300293

[18] Clements JS, Sato M, Davis RH. Preliminary investigation of prebreakdown phenomena and chemical reactions using a pulsed high-voltage discharge in water. IEEE Transactions on Industry Applications. 1987;**IA-23**:224-235. DOI: 10.1109/TIA.1987.4504897

[19] Bruggeman P, Leys C. Non-thermal plasmas in and in contact with liquids. Journal of Physics D: Applied Physics. 2009;**42**:053001. DOI: 10.1088/0022-3727/42/5/053001

[20] Takahashi K, Sasaki Y, Mukaigawa S, Takaki K, Fujiwara T, Satta N. Purification of high-conductivity water using gas–liquid phase discharge reactor. IEEE Transactions on Plasma Science. 2010;**38**:2694-2700. DOI: 0.1109/TPS.2010.2049505

[21] Levko D, Sharma A, Raja LL. Plasmas generated in bubbles immersed in liquids: direct current streamers versus microwave plasma. Journal of Physics D: Applied Physics. 2016;**49**: 285205. DOI: 10.1088/0022-3727/49/28/285205

[22] Hamdan A, Cha MS. The effects of gaseous bubble composition and gap distance on the characteristics of nanosecond discharges in distilled water. Journal of Physics D: Applied Physics. 2016;**49**:245203. DOI: 10.1088/0022-3727/49/24/245203

[23] Bruggeman PJ, Kushner MJ, Locke BR, Gardeniers JGE, Graham WG, Graves DB, et al. Plasma–liquid interactions: a review and roadmap. Plasma Sources Science and Technology. 2016;**25**:53002. DOI: 10.1088/0963-0252/25/5/053002

[24] Malik MA. Water purification by plasma which reactors are most energy efficient? Plasma Chemistry and Plasma Processing. 2010;**30**:21-31. DOI: 10.1007/s11090-009-9202-2

[25] Kanazawa S, Kawano H, Watanabe S, Furuki T, Akamine S, Ichiki R, et al. Observation of OH radicals produced by pulsed discharges on the surface of a liquid. Plasma Sources Science and Technology. 2011; **20**:034010. DOI: 10.1088/0963-0252/20/3/034010

[26] Hoffer P, Sugiyama Y, Hosseini SHR, Akiyama H, Lukes P, Akiyama M. Characteristics of meter-scale surface electrical discharge propagating along water surface at atmospheric pressure. Journal of Physics D: Applied Physics. 2016;**49**:415202. DOI: 10.1088/0022-3727/49/41/415202

[27] Xiong Y, Zhang Q, Wandell R, Bresch S, Wang H, Locke BR. Synergistic 1 , 4-dioxane removal by non-thermal plasma followed by biodegradation Synergistic 1,4-dioxane removal by non-thermal plasma followed by biodegradation. Chemical

Engineering Journal. 2018;**361**:519-527. DOI: 10.1016/j.cej.2018.12.094

[28] Wandell RJ, Wang H, Bulusu RKM, Gallan RO, Locke BR. Formation of nitrogen oxides by nanosecond pulsed plasma discharges in gas – Liquid reactors. Plasma Chemistry and Plasma Processing. 2019;**39**:643-666. DOI: 10.1007/s11090-019-09981-w

[29] Takahashi K, Konno R, Satta N, Takaki K. Estimation of hydroxyl radicals produced by pulsed discharge inside bubble in water using indigo carmine as chemical probe. IEEE Transactions on Plasma Science. 2018;**46**:663-669. DOI: 10.1109/TPS.2018.2792480

[30] Li J, Sato M, Ohshima T. Degradation of phenol in water using a gas–liquid phase pulsed discharge plasma reactor. Thin Solid Films. 2007; **515**:4283-4288. DOI: 10.1016/j.tsf.2006.02.070

[31] Takahashi K, Yagi I, Takaki K, Satta N. Development of pulsed discharge inside bubble in water. IEEE Transactions on Plasma Science. 2011; **39**:2654-2655. DOI: 10.1109/TPS.2011.2164095

[32] Takahashi K, Takayama H, Kobayashi S, Takeda M, Nagata Y, Karashima K, et al. Observation of the development of pulsed discharge inside a bubble under water using ICCD camera. Vacuum. 2020;**182**:109690. DOI: 10.1016/j.vacuum.2020.109690

[33] Sharma A, Levko D, Raja LL, Cha MS. Kinetics and dynamics of nanosecond streamer discharge in atmospheric-pressure gas bubble suspended in distilled water under saturated vapor pressure conditions. Journal of Physics D: Applied Physics. 2016;**49**:395205. DOI: 10.1088/0022-3727/49/39/395205

[34] Tachibana K, Takekata Y, Mizumoto Y, Motomura H, Jinno M. Analysis of a pulsed discharge within single bubbles in water under synchronized conditions. Plasma Sources Science and Technology. 2011; **20**:34005. DOI: 10.1088/0963-0252/20/3/034005

[35] Physics A, Sobota A, Universiteit T, Technische V, Eindhoven U, Philips M. Speed of streamers in argon over a flat surface of a dielectric. Journal of Physics D: Applied Physics. 2008;**42**:015211. DOI: 10.1088/0022-3727/42/1/015211

[36] Babaeva NY, Tereshonok DV, Naidis GV. Fluid and hybrid modeling of nanosecond surface discharges: effect of polarity and secondary electrons emission. Plasma Sources Science and Technology. 2016;**25**:044008. DOI: 10.1088/0963-0252/25/4/044008

[37] Bruggeman P, Guns P, Degroote J, Vierendeels J. Influence of the water surface on the glow-to-spark transition in a metal-pin-to-water electrode system. Plasma Sources Science and Technology. 2008;**17**:045014. DOI: 10.1088/0963-0252/17/4/045014

[38] Takahashi K, Takeda M, Konno R, Takaki K, Satta N. Influence of electric parameters on hydroxyl radical production by positive pulsed discharge inside of a bubble in water. IEEE Transactions on Plasma Science. 2019; **47**:1105-1113. DOI: 10.1109/TPS.2018.2883767

[39] Takahashi K, Mukaigawa S, Takaki K, Fujiwara T, Satta N. Water purification using non-thermal plasma driven by blumlein-line stacked pulsed power generator. Journal of Plasma and Fusion Research SERIES. 2009;**8**:1459-1462

[40] Minamitani Y, Yamada T. Investigation of the influence of droplets to streamer discharge in water treatment by pulsed discharge in air spraying. IEEE Transactions on Plasma Science. 2016;**44**:2173-2180. DOI: 10.1109/TPS.2016.2592515

[41] Stefanovic I, Bibinov N, Deryugin A, Vinoradov I, Napartovich A, Wiesemann K. Kinetics of ozone and nitric oxides in dielectric barrier discharges in O2/NOx and N2/O2/NOx mixtures. Plasma Sources Science and Technology. 2001;**10**: 406-416. DOI: 10.1088/0963-0252/10/3/303

[42] Luo Y, Lietz AM, Yatom S, Kushner MJ, Bruggeman PJ. Plasma kinetics in a nanosecond pulsed filamentary discharge sustained in Ar – H2O and H2O. Journal of Physics D: Applied Physics. 2019;**52**:044003. DOI: 10.1088/1361-6463/aaeb14

[43] Liu Y, Tan Z, Chen X, Li X, Wang X. A numerical investigation on the effects of water vapor on electron energy and OH production in atmospheric-pressure He/H2O and Ar/H2O plasma jets. IEEE Transactions on Plasma Science. 2019;**47**:1593-1604. DOI: 10.1109/TPS.2019.2896060

[44] Shirafuji T, Murakami T. Contribution of electrons, Ar(3P0,2), H2O+, and H3O+ to production of OH (A2Σ+) in a micro-dielectric barrier discharge of Ar/H2O. Japanese Journal of Applied Physics. 2014;**54**:01AC03. DOI: 10.7567/JJAP.54.01AC03

[45] Verreycken T, Schram DC, Leys C, Bruggeman P. Spectroscopic study of an atmospheric pressure dc glow discharge with a water electrode in atomic and molecular gases. Plasma Sources Science and Technology. 2010;**19**:045004. DOI: 10.1088/0963-0252/19/4/045004

[46] Komuro A, Ono R, Oda T. Behaviour of OH radicals in an atmospheric-pressure streamer discharge studied by two-dimensional numerical simulation. Journal of Physics D: Applied Physics. 2013;**46**:175206. DOI: 10.1088/0022-3727/46/17/175206

[47] Bruggeman P, Liu J, Degroote J, Kong MG. DC excited glow discharges in atmospheric pressure air in pin-to-water electrode. Journal of Physics D: Applied Physics. 2008;**41**:215201. DOI: 10.1088/0022-3727/41/21/215201

[48] Wakisaka S, Tsuda K, Takahashi K, Satoh K. Mechanism of pH variation and H 2 O 2 generation in water exposed to pulsed discharge plasma. IEEE Transactions on Plasma Science. 2019; **47**:1083-1088. DOI: 10.1109/TPS.2018.2866282

[49] Atkinson R, Baulch DL, Cox RA, Hampson RF, Kerr JA, Troe J. Evaluated kinetic and photochemical data for atmospheric chemistry: Supplement III. IUPAC subcommittee on gas kinetic data evaluation for atmospheric chemistry. Journal of Physical and Chemical Reference Data. 1989;**18**:881. DOI: 10.1063/1.555832

[50] Gerasimov GY, Shatalov OP. Kinetic mechanism of combustion of hydrogen–oxygen mixtures. Journal of Engineering Physics and Thermophysics. 2013;**86**: 987-995. DOI: 10.1007/s10891-013-0919-7

[51] Yagi I, Ono R, Oda T, Takaki K. Two-dimensional LIF measurements of humidity and OH density resulting from evaporated water from a wet surface in plasma for medical use. Plasma Sources Science and Technology. 2014;**24**: 015002. DOI: 10.1088/0963-0252/24/1/015002

[52] Takeuchi N, Ando M, Yasuoka K. Investigation of the loss mechanisms of hydroxyl radicals in the decomposition of organic compounds using plasma generated over water. Japanese Journal of Applied Physics. 2015;**54**:116201. DOI: 10.7567/JJAP.54.116201

[53] Chen C, Liu DX, Liu ZC, Yang AJ, Chen HL, Shama G, et al. A model of plasma-biofilm and plasma-tissue interactions at ambient pressure. Plasma Chemistry and Plasma Processing. 2014;

34:403-441. DOI: 10.1007/s11090-014-9545-1

[54] Rumbach P, Bartels DM, Go DB. The penetration and concentration of solvated electrons and hydroxyl radicals at a plasma-liquid interface. Plasma Sources Science and Technology. 2018; **27**:115013. DOI: 10.1088/1361-6595/aaed07

[55] Kanazawa S, Kawano H, Watanabe S, Furuki T, Akamine S, Ichiki R, et al. Observation of OH radicals produced by pulsed discharges on the surface of a liquid. Plasma Sources Science and Technology. 2011; **20**:34010. DOI: 10.1088/0963-0252/20/3/034010

[56] Gomes A, Fernandes E, Lima JLFC. Fluorescence probes used for detection of reactive oxygen species. Journal of Biochemical and Biophysical Methods. 2005;**65**:45-80. DOI: 10.1016/j.jbbm.2005.10.003

[57] Shiraki D, Ishibashi N, Takeuchi N. Quantitative estimation of OH radicals reacting in liquid using a chemical probe for plasma in contact with liquid. IEEE Transactions on Plasma Science. 2016; **44**:3158-3163. DOI: 10.1109/TPS.2016.2603157

[58] Knipping EM, Dabdub D. Modeling Cl2 formation from aqueous NaCl particles: Evidence for interfacial reactions and importance of Cl2 decomposition in alkaline solution. Journal of Geophysical Research. 2002; **107**:4360. DOI: 10.1029/2001JD000867

[59] Yuchen H, Uehara S, Tanaka H, Nishiyama H. Numerical modelling and simulation of chemical reactions in a nano-pulse discharged bubble for water treatment. Plasma Science and Technology. 2016;**18**:924-934. DOI: 10.1088/1009-0630/18/9/09

[60] Ono R, Oda T. Dynamics of ozone and OH radicals generated by pulsed corona discharge in humid-air flow reactor measured by laser spectroscopy. Journal of Applied Physics. 2003;**93**: 5876-5882. DOI: 10.1063/1.1567796

[61] Nakagawa Y, Ono R, Oda T. Effect of discharge polarity on OH density and temperature in coaxial-cylinder barrier discharge under atmospheric pressure humid air. Japanese Journal of Applied Physics. 2018;**57**:096103. DOI: 10.7567/JJAP.57.096103

[62] Staehelln J, Holgne J. Decomposition of ozone in water: Rate of initiation by hydroxide ions and hydrogen peroxide. Environmental Science & Technology. 1982;**16**:676-681. DOI: 10.1021/es00104a009

[63] Kim HH, Prieto G, Takashima K, Katsura S, Mizuno A. Performance evaluation of discharge plasma process for gaseous pollutant removal. Journal of Electrostatics. 2002;**55**:25-41. DOI: 10.1016/S0304-3886(01)00182-6

[64] Soloshenko IA, Tsiolko VV, Pogulay SS, Kalyuzhnaya AG, Bazhenov VY, Shchedrin AI. Effect of water adding on kinetics of barrier discharge in air. Plasma Sources Science and Technology. 2009;**18**:045019. DOI: 10.1088/0963-0252/18/4/045019

[65] Machala Z, Tarabova B, Hensel K, Spetolikova E, Sikurova L, Lukes P. Formation of ROS and RNS in water electro-sprayed through transient spark discharge in air and their bactericidal effects. Plasma Processes and Polymers. 2013;**10**:649-659. DOI: 10.1002/ppap.201200113

[66] Régimbal JM, Mozurkewich M. Peroxynitric acid decay mechanisms and kinetics at low pH. The Journal of Physical Chemistry. A. 1997;**101**: 8822-8829. DOI: 10.1021/jp971908n

[67] Warneck P. Chemical reactions in clouds. Fresenius' Journal of Analytical

Chemistry. 1991;**340**:585. DOI: 10.1007/BF00322434

[68] Hoigne J, Bader H, Haag WR, Staehelin J. Rate constants of reactions of ozone with organic and inorganic compounds in water—III. Inorganic compounds and radicals. Water Research. 1985;**19**:993-1004. DOI: 10.1016/0043-1354(85)90368-9

[69] Takahashi K, Kawamura S, Takada R, Takaki K, Satta N, Fujio T. Influence of a plasma-treated nutrient solution containing 2,4-dichlorobenzoic acid on the growth of cucumber in a hydroponic system. Journal of Applied Physics. 2021;**129**:143301. DOI: 10.1063/5.0044328

[70] Hug SJ, Leupin O. Iron-catalyzed oxidation of Arsenic(III) by oxygen and by hydrogen perxide: pH-dependent formation of oxidants in the fenton reaction. Environmental Science & Technology. 2003;**37**:2734-2742. DOI: 10.1021/es026208x

[71] Kremer ML. Mechanism of the Fenton reaction. Evidence for a new intermediate. Physical Chemistry Chemical Physics. 1999;**1**:3595-3605. DOI: 10.1039/A903915E

[72] White EM, Vaughan PP, Zepp RG. Role of the photo-Fenton reaction in the production of hydroxyl radicals and photobleaching of colored dissolved organic matter in a coastal river of the southeastern United States. Aquatic Sciences. 2003;**65**:402-412. DOI: 10.1007/s00027-003-0675-4

[73] Takahashi K, Takaki K, Satta N. Water remediation using pulsed power discharge under water with advanced oxidation process. Journal of Advanced Oxidation Technologies. 2012;**15**:365-373. DOI: 10.1515/jaots-2012-0216

[74] Liu Q, Schurter LM, Muller CE, Aloisio S, Francisco JS, Margerum DW. Kinetics and mechanisms of aqueous ozone reactions with bromide, sulfite, hydrogen sulfite, iodide, and nitrite ions. Inorganic Chemistry. 2001;**40**:4436-4442. DOI: 10.1021/ic000919j

[75] Buxton GV, Greenstock CL, Helman WP, Ross AB. Critical review of rate constants for reactions of hydrated electrons, hydrogen atoms and hydroxyl radicals in aqueous solution. Journal of Physical and Chemical Reference Data. 1987;**17**:513. DOI: 10.1063/1.555805

[76] Ross F, Ross AB. Selected specific rates of reactions of transients from water in aqueous solution. In: III. Hydroxyl Radical and Perhydroxyl Radical and Their Radical Ions. United States: N: Web; 1977. DOI: 10.2172/6635906

[77] Logan SR. Redox reactions of organic radicals with ferrocene/ferricenium species in aqueous solution. Part 1. Radicals derived from carboxylic acids. Journal of the Chemical Society, Perkin Transactions. 1989;**2**:751-754. DOI: 10.1139/v91-081

[78] Hoigne J, Bader H. Rate constants of reactions of ozone with organic and inorganic compounds in waterII Dissociating organic compounds. Water Research. 1983;**17**:158-194. DOI: 10.1016/0043-1354(83)90098-2

[79] Neta P, Huie RE, Ross AB. Rate constants for reactions of inorganic radicals in aqueous solution rate · constants for reactions of inorganic radicals in aqueous solution. Journal of Physical and Chemical Reference Data. 2009;**17**:1027-1284. DOI: 10.1063/1.555808

[80] Abe A. Distribution of 1, 4-dioxane in relation to possible sources in the water environment. Science of the Total Environment. 1999;**227**:41-47. DOI: 10.1016/S0048-9697(99)00003-0

[81] Zenker MJ, Borden RC, Barlaz MA. Occurrence and treatment of

1,4-dioxane in aqueous environments. Environmental Engineering Science. 2003;**20**:423-443. DOI: 10.1089/109287503768335913

[82] Khan FI, Ghoshal AK. Removal of volatile organic compounds from polluted air. Journal of Loss Prevention in the Process Industries. 2000;**13**: 527-545. DOI: 10.1016/S0950-4230(00)00007-3

[83] Huang B, Lei C, Wei C, Zeng G. Chlorinated volatile organic compounds (Cl-VOCs) in environment — sources, potential human health impacts, and current remediation technologies. Environment International. 2014;**71**: 118-138. DOI: 10.1016/j.envint.2014.06.013

[84] Bian W, Song X, Liu D, Zhang J, Chen X. Actions of nitrogen plasma in the 4-chrolophenol degradation by pulsed high-voltage discharge with bubbling gas. Chemical Engineering Journal. 2013;**219**:385-394. DOI: 10.1016/j.cej.2012.12.074

[85] Nunnally T, Tsangaris A, Rabinovich A, Nirenberg G. Gliding arc plasma oxidative steam reforming of a simulated syngas containing naphthalene and toluene. International Journal of Hydrogen Energy. 2014;**39**: 11976-11989. DOI: 10.1016/j.ijhydene.2014.06.005

[86] Yang J, Zeng D, Hassan M, Ma Z, Dong L, Xie Y, et al. Efficient degradation of Bisphenol A by dielectric barrier discharge non-thermal plasma: Performance, degradation pathways and mechanistic consideration. Chemosphere. 2021;**286**:131627. DOI: 10.1016/j.chemosphere.2021.131627

[87] Lei W, Huifen Z, Xin Y. Dechlorination and decomposition of trichloroacetic acid by glow discharge plasma in aqueous solution. Electrochimica Acta. 2014;**115**:332-336. DOI: 10.1016/j.electacta.2013.10.160

[88] Ghodbane H, Hamdaoui O, Vandamme J, Durme JV, Vanraes P, Leys C, et al. Degradation of AB25 dye in liquid medium by atmospheric pressure non-thermal plasma and plasma combination with photocatalyst TiO2. Open Chemistry. 2014;**13**:325-331. DOI: 10.1515/chem-2015-0040

[89] Shirafuji T, Ishida Y, Nomura A, Hayashi Y, Goto M. Reaction mechanisms of methylene-blue degradation in three-dimensionally integrated micro-solution plasma. Jpn. Journal of Applied Physics. 2017;**56**: 06HF02. DOI: 10.7567/JJAP.56.06HF02

[90] Iwabuchi M, Takahashi K, Takaki K, Satta N. Influence of sodium carbonate on decomposition of formic acid by pulsed discharge plasma inside bubble in water. Japanese Journal of Applied Physics. 2016;**55**:07LF02. DOI: 10.7567/JJAP.55.07LF02

[91] Takahashi K, Takayama H, Yagi I, Takaki K, Satta N. Decomposition process of volatile organic compounds dissolved into water by pulsed discharge inside bubble. Japanese Journal of Applied Physics. 2020;**59**:SHHA06. DOI: 10.35848/1347-4065/ab72cd

[92] Reife A, Freeman HS. Environmental Chemistry of Dyes and Pigments. New York, NY: John Wiley & Sons; 1996. pp. 50-59

[93] Backett MA, Hua I. Elucidation of the 1,4-dioxane decomposition pathway at discrete ultrasonic frequencies. Environmental Science & Technology. 2000;**34**:3944-3953. DOI: doi.org/10.1021/es000928r

[94] Stefan MI, Bolton JR. Mechanism of the degradation of 1,4-dioxane in dilute aqueous solution using the UV/hydrogen peroxide process. Environmental Science & Technology. 1998;**32**:1588-1595. DOI: 10.1021/es970633m

[95] Höbel B, Sonntag C. OH-Radical induced degradation of ethylenediaminetetraacetic acid(EDTA) in aqueous solution; a pulse raiolysis study. Journal of the Chemical Society, Perkin Transactions. 1998;**2**:509-513. DOI: 10.1039/A708167G

[96] Thomas JK. Rates of reaction of the hydroxyl radical. Transactions of the Faraday Society. 1966;**61**:702. DOI: 10.1039/TF9656100702

[97] Haag WR, Yao CCCD. Rate constants for reaction off hydroxyl radicals with several drinking water contaminants. Environmental Science & Technology. 1992;**26**:1005-1013. DOI: 10.1021/es00029a021

[98] Fitzsimmons C, Ismail F, Whitehead JC, Wilman JJ. The chemistry of dichloromethane destruction in atmospheric-pressure gas streams by a dielectric packed-bed plasma reactor. The Journal of Physical Chemistry. A. 2000;**104**:6032-6038. DOI: doi.org/10.1021/jp000354c

[99] Kundu SK, Kennedy EM, Mackie JC, Holdsworth CI, Molloy TS, Gaikwad VV, et al. Study on the reaction of CCl2F2 with CH4 in a dielectric barrier discharge nonequilibrium plasma. Plasma Processes and Polymers. 2013;**10**: 912-921. DOI: 10.1002/ppap.201300053

[100] Gaikwad V, Kennedy E, Mackie J, Holdsworth C, Molloy S, Kundu S, et al. Reaction of dichloromethane under non-oxidative conditions in a dielectric barrier discharge reactor and characterisation of the resultant polymer. Chemical Engineering Journal. 2016;**290**:499-506. DOI: 10.1016/j.cej.2015.12.105

[101] Orlandini I, Riedel U. Chemical kinetics of NO removal by pulsed corona discharges. Journal of Physics D: Applied Physics. 2000;**33**:2467

[102] Takaki K, Takahashi K, Hamanaka D, Yoshida R, Uchino T. Function of plasma and electrostatics for keeping quality of agricultural produce in post-harvest stage. Japanese Journal of Applied Physics. 2021;**60**: 010501. DOI: 10.35848/1347-4065/abcc13

[103] Takaki K, Hayashi N, Wang D, Ohshima T. High-voltage technologies for agriculture and food processing. Journal of Physics D: Applied Physics. 2019;**52**:473001. DOI: 10.1088/1361-6463/ab2e2d

[104] Puac N, Gherardi M, Shiratani M. Plasma agriculture : A rapidly emerging field. Plasma Processes and Polymers. 2018;**15**:1700174. DOI: 10.1002/ppap.201700174

[105] Ranieri P, Kizer J, Gatiboni L, Hernández R, Pierce MR, Stapelmann K, et al. Plasma agriculture : Review from the perspective of the plant and its ecosystem. Plasma Processes and Polymers. 2020;**18**:e2000162. DOI: 10.1002/ppap.202000162

[106] Okumura T, Saito Y, Takano K, Takahashi K, Takaki K, Satta N, et al. Inactivation of bacteria using discharge plasma under liquid fertilizer in a hydroponic culture system. Plasma Medicine. 2017;**6**:247-254. DOI: 10.1615/PlasmaMed.2016018683

[107] Takahata J, Takaki K, Satta N, Takahashi K. Improvement of growth rate of plants by bubble discharge in water. Japanese Journal of Applied Physics. 2015;**54**:01AG07. DOI: 10.7567/JJAP.54.01AG07

[108] Takahashi K, Saito Y, Oikawa R, Okumura T, Takaki K, Fujio T. Development of automatically controlled corona plasma system for inactivation of pathogen in hydroponic cultivation medium of tomato. Journal of Electrostatics. 2018;**91**:61-69. DOI: 10.1016/j.elstat.2017.12.006

[109] Schnabel U, Andrasch M, Weltmann K. Antimicrobial efficiency of non-thermal atmospheric pressure plasma processed water (PPW) against agricultural relevant bacteria suspensions. International Journal of Environmental & Agriculture Research. 2016;**2**:212-224

[110] Ma R, Wang G, Tian Y, Wang K, Zhang J, Fang J. Non-thermal plasma-activated water inactivation of food-borne pathogen on fresh produce. Journal of Hazardous Materials. 2015; **300**:643-651. DOI: 10.1016/j.jhazmat.2015.07.061

[111] Xu Y, Tian Y, Ma R, Liu Q, Zhang J. Effect of plasma activated water on the postharvest quality of button mushrooms, Agaricus bisporus. Food Chemistry. 2016;**197**:436-444. DOI: 10.1016/j.foodchem.2015.10.144

[112] Andrasch M, Stachowiak J, Schlüter O, Schnabel U, Ehlbeck J. Scale-up to pilot plant dimensions of plasma processed water generation for fresh-cut lettuce treatment. Food Packaging and Shelf Life. 2017;**14**:40-45. DOI: 10.1016/j.fpsl.2017.08.007

Chapter 2

Irradiation of Sewage Sludge

Roberto M. Uribe and Jean Engohang-Ndong

Abstract

A review of the current status of sewage sludge decontamination using electron beam irradiation at industrial scale is presented. The chapter includes a historical development of the technology using both gamma and electron beam sources, a description of a facility using an electron accelerator, a discussion of the quality control techniques used to certify that satisfactory decontamination levels for safe use of treated sludge have been achieved, the effect of electron beam irradiation on the bacteria and virus present in a typical sample of municipal sewage sludge, and an analysis of the costs of decontaminating sewage sludge using electron beam irradiation compared to traditional and more routine technologies. Finally, the chapter concludes by emphasizing on the fact that electron accelerators described in this chapter are capable to decontaminate a typical municipal sewage sludge at competitive costs which are shown to be comparable and/or lower than routinely used technologies to achieve class A biosolids by the Environmental Protection Agency standards.

Keywords: sludge decontamination, e-beam applications, microbial pathogens

1. Introduction

The use of radiation to reduce the microbial contamination in foodstuffs and medical supplies is a well-established technology. Both gamma rays and high energy electrons are used with this purpose and their main effect is to damage the DNA of harmful microorganisms. Although the technology has also been used in environmental applications like the removal of contaminants from fume stacks and the decontamination of wastewater and sewage sludge, their application at a commercial stage has not been as successful as in the case of medical supplies or food irradiation.

This chapter presents first an historical overview of the area of sewage sludge irradiation and the technologies developed over time. Then a description of the technologies used for the electron beam treatment of sewage sludge will be presented. A typical facility that has been developed to irradiate sewage sludge with electrons will be described as well as a current state of the use of the technology in several places around the world.

The techniques used to assess the quality control of the process will be described including dosimetric techniques and the analysis of the effect of the electron beam on the reduction of microorganisms that contaminate sewage sludge.

2. Historical development of the technology

Sludges from municipal sewage systems are good soil fertilizer because of their high content of organic matter, nitrogen, phosphorus, and many trace metals essential to plants pointing to its potential in agricultural applications, however they also contain a large density of pathogenic microorganisms, parasites and parasitic eggs that cause diseases to human beings, pets, and farm animals. To make the use of sewage sludge as a safe valuable soil fertilizer in agricultural applications it needs to be disinfected. To that end, several approaches have been used over the years including old techniques such as incineration and more modern techniques such as irradiation.

The use of irradiation to disinfect sludge started in 1973 when an industrial gamma ray facility from Geiselbullach near Munich (Germany) used Co-60 and Cs-137 sources [1]. The facility used 90,000 Ci of Co-60 and 570,000 Ci of Cs-137 and treated up to 180 m^3/day of sludge. Similar activities were undertaken in the United States by the US Department of Energy Sandia National Laboratories using Cs-137 and capable of operating up to a maximum capacity of 7250 kg/day [2].

The use of electron accelerators to disinfect sludge was also started in the 1970's with the work by Trump and collaborators [3] in Cambridge Massachussets, USA. The system which was installed at the Deer Island Wastewater Treatment Plant in Boston Massachussets originally in 1976 consisted of a 50 kW High Voltage Engineering (HVE) electron accelerator and originally delivered up to 375,000 l/day (100,000 gpd) of sludge irradiated at a dose of 4 kGy. In 1980 the system was completely restructured to increase the capacity of the plant to 637,500 l/day (170,000 gpd). In 1982 a similar system was installed in Miami, Florida with a 1.5 MeV, 50 mA accelerator used to treat sludge at a throughput of 645 m^3/day [4]. Aqueous streams were presented to the beam in a falling stream about 114 cm wide and 0.4 cm thick. The system was capable to irradiate water streams to doses up to 8 kGy by changing the beam current from 0 to 50 mA [5]. A similar system was installed in Brazil where a 1.5 MeV, 37.5 kW accelerator, with a maximum throughput of 45 l/min were described [4]. Chmielewski and collaborators [6] have described the activities developed in Poland in this area including feasibility studies on the technique and then the design of a 70 tons/day treatment plant for doses of 5–6 kGy using a 300 kW, 10 MeV electron accelerator.

Similar studies have been conducted in Japan since the 1970's by the Japan Atomic Energy Research Institute (JAERI). In this respect Washino [7] has compared electron and gamma ray treatment of wastewater in order to determine their bactericidal effect and found a larger reduction in the concentration of microorganisms for the latter one, and mention that this is due to an oxygen destruction effect produced by the higher dose rate generally used in the case of the electron irradiation. To overcome this problem, he proposed a reaction chamber for the irradiation with electrons consisting of a concentric dual-tube bubbling column with a 50 micron thick stainless steel window at the top to allow for the entrance of electrons. Oxygen is bubbling from the bottom of the reaction chamber to compensate for this reduction effect. Similarly, Hashimoto and co-workers [8] have described the use of process-control techniques to make the electron irradiation of waste waters more effective.

In 2014 through a collaboration between Arlington County in Virginia, USA and Kent State University, in Kent Ohio, USA a sample of sludge was irradiated to demonstrate the feasibility and the economic value of the process. The sample consisting of 33,750 l (9,000 gal) of sludge was irradiated at 6.7 and 25.7 kGy under typical production conditions using 3 MeV electrons provided by a Dynamitron accelerator and a flow rate of 184 l/min (50 gallons/min) and showed that the process is effective in reducing the concentration of some microorganisms and more economical than conventional disinfection techniques [9].

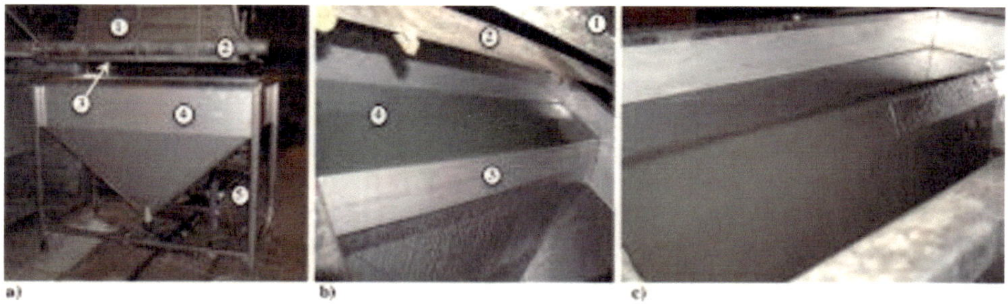

Figure 1.
Example of a delivery system to irradiate sludge with high energy electrons. Photo (a) shows the entire system including the scanner of the accelerator (1), the air blower (2) to refrigerate the titanium window (3), the sludge delivery and collection tank (4), and the piping system to introduce the sludge in the tank from the bottom (5). The photo in (b) shows the scanner of the electron accelerator (1), the accelerator shutter (2), the weird with the irradiated sludge, simulated with water (3), and the incoming sludge (4). Photo in (c) shows an actual "curtain" of sludge flowing through the system.

3. Description of a typical electron beam treatment facility

Usually, sludge consists of wastewater with a low concentration of solid particles (around 15%). This is a fluid that can readily move from the storage area to the irradiation zone and back to the storage area and in a typical sludge irradiation facility this is done using a hydraulic system consisting of pipes and pneumatic pumps. Inside the electron beam room, the sludge is irradiated using a bulk irradiation system that will present a laminar like fluid to the electron beam with a thickness of a few millimeters depending on the energy of the electrons. An example of this could be a rectangular stainless-steel tank divided in the middle by a wall. At the top of this wall, a weir is built to produce a cascade of sludge. This weir is aligned with the length of the scanner of the accelerator in such manner that the sludge is uniformly irradiated by the system. The influent (pre-treated) sludge flows from the bottom of the first section of the tank, rising up to the height of the middle wall over the weir and then overflowing into the second section (**Figure 1** shows an example of this system, as installed for the Arlington County experiment described above). The effluent (post-treated) sludge is taken out from the bottom of the second section flowing through a second set of rubber pipes to a second storage tank.

4. Quality control of the process

The effect of radiation on the reduction of bacterial load and decontamination of sludge depends on the amount of energy from the radiation that is absorbed by the sludge. This energy is used to produce chemically active species that eventually disrupt structural integrity of DNA in bacteria, parasites, and viruses causing their partial or permanent inactivation and eventually death of microorganisms. The amount of energy from the radiation that is absorbed by the sludge is measured by the physical term dose which is defined as the amount of energy absorbed in a volume of sludge divided by the mass of that volume and is measured in kilogray (kGy). A dose of several kGy could be enough to cause a disruption of DNA molecules and inactivate a virus or kill a bacterium. In order to apply the technique in a municipal installation one needs to be sure that all the sludge that gets irradiated will receive the minimum amount of dose needed to inactivate the microorganisms. Several techniques have been proposed to determine the real dose received by the sample during the irradiation of sludge. One of them makes use of the fact that the interaction of the radiation with the sludge will cause a certain amount of temperature increase in the sludge, therefore, the dose can be determined by measuring the difference in temperature in it before and after irradiation using thermocouples located in the influent and effluent parts of the water piping system near the falling stream; this set of thermocouples can also be used to monitor the irradiation treatment. An advantage of this technique is that once the thermocouples are installed the temperature difference between the input and output portions of the system is easy to be determined; the only two parameters needed to know are the specific capacity of the sample which needs to be determined experimentally for each type of sludge treated and the temperature difference. With these two values the energy absorbed in a sample of sludge can readily be obtained. One problem with this technique however is that it only provides an average value of the dose and does not measure the dose that individual samples of sludge will get when going through the system neither does it measure how uniform the dose delivered to the sludge is really. To overcome the first problem, a small dosimeter can typically be placed inside a small water tight plastic capsule and be allowed to run through the system from the supply tank to the collection tank. Several candidates of dosimeters can be used in this respect; one of them consists of alanine pellets which have dimensions of several mm. These dosimeters rely on the fact that under irradiation a concentration of relatively stable free radicals are produced in them; the free concentration of free radicals can be determined by the technique of electron paramagnetic resonance [9, 10]. Another dosimeter that can be used is Lithium fluoride in crystals or in powder; the crystals have dimensions of a few milimeters per side and two or three could be accommodated in a single capsule. When using it in powder a few mg can be placed in a capsule for irradiation. Upon irradiation electrons from the valence band of the crystal jump to the conduction band and when trying to be de-excited, some of these electrons get trapped in the energy gap of the material. The electrons can be liberated from these traps and allowed to return to the valence band by applying heat to the material. When the electrons return to the valence band, they emit light to lose the excess of energy they have in them. The technique to heat the crystals in a controlled way and measure the output light in the de-excitation process is called thermoluminescence and when applied to dosimetry it is called thermoluminescence dosimetry (TLD) [11]. Uribe and co-workers described the use of this technology in a similar application in the irradiation of corn kernels in a pilot plant using 1.0 MeV electrons [12].

To address the second issue and as part of the operation qualification (OQ) activities of the system [13] a verification of the dose uniformity along the weird where the electrons irradiate the sludge needs to be performed. This is useful to remove "dark" spots where the sludge does not get irradiated or to correct for instabilities in the electron beam scanning system that makes the electrons to stay longer times or to correct for misalignment of the sludge delivery system with the scanned electrons coming out of the electron accelerator. Film systems are the best candidates to perform dose uniformity measurements along the scanner of the electron accelerator. Several examples of these are available in the market; good examples of them are the cellulose triacetate film (CTA) [14] and the radio chromic films [15]. Both of these work on the principle that radiation induces a change in the optical absorption of the film that can be quantified using a spectrophotometer. Through a suitable calibration with a primary dosimeter these systems can be used to measure the dose along the length of the weird of the sludge delivery system. For instance, when performing the OQ activities using CTA film it was found that the length of the weird where the sludge was irradiated was longer than the extent of the scanned electrons. So, there were "dark" spots at both ends of the weird where the sludge was not irradiated. The situation was corrected by placing a couple of plates at both ends of the weird that reduced its length so all the sludge going through the weird was irradiated [9].

5. Bactericidal effect

Until a recent past, the radiation effects on the bacterial load and removal of noxious chemical compounds had only been performed in small samples of sludge irradiated under laboratory conditions and mainly address either only the microbiological or the chemical effect of radiation in a sample of sludge [16–20]. Processing and disposal of wastewater sludge is a critical problem worldwide [18] and especially in large metropolitan such as Washington DC in the USA and Tokyo in Japan. Therefore, new technologies to solve the problem of safely disposing of sewage sludge are constantly being sought. An attractive solution for the disposal of wastewater sludge is its utilization in crop fields and landscaping as a fertilizer due to its high content in natural nitrogen compounds. However, in order to be used as such, it must be converted into a class A biosolid [21] which is a form of treated sludge that is deemed safe for humans and animals. In recent years, electron beam technology have shown to be an economically alternative that could be used to meet these regulations by considerably reducing the number of potentially harmful bacteria such as the fecal coliform bacteria and helminth ova such as *Ascaris ova* which are both ubiquitous contaminants present in the sewage sludge. The fecal coliform bacteria are a group of bacteria that are released in the environment through the fecal excrement of humans, wildlife, and livestock. They typically occur in the digestive tract of humans and warm-blooded animals. When present in sewage sludge, they are indicative of the presence of human and animal pathogens including some strains of *Escherichia coli*, *Shigella flexneri*, and *Salmonella typhimurium*. Among the fecal coliforms, the *E. coli O157:H7* strain which was responsible for many outbreaks has retain attention of media in the past several years. Indeed, *E. coli O157:H7* was found contaminating drinking water and vegetables, causing cases of stomach cramping, vomiting, and bloody diarrhea in patients. However, in most cases infections with the *E. coli O157:H7* strain were mild or with no symptoms. Most people infected with *S. typhimurium* develop diarrhea, stomach cramping, and fever which could cause the patient to experience fatigues and dehydration. An infection with *S. flexneri* is much more serious with severe stomach pain, dehydration, severe diarrhea, and fever. Patients suffering for shigellosis feel very sick and stay in bed at the pick of the infection. In farming areas where livestock is grown, beside occurrence of fecal coliforms, ascaris ova that when ingested by a person or an animal would hatch and develop into adult worms, *Ascaris lumbricoides* cause stomach pain, nausea, vomiting, and fatigue. In patients who are heavily infested with ascaris worms, the parasites could be expelled through feces and/or vomit.

Additionally, the electron beam technology could be used to reduce the concentration of volatile organic sulfides and other volatile organic compounds (VOCs) responsible for the unpleasant odor in sewage sludge. Moreover, electron beam treatment of sludge is not an energy intensive process which means that it has a small footprint and the processing times are usually short. The main effect of radiation on sewage sludge is in the radiolysis of water producing OH^* and H^* radicals and hydrated electrons, highly reactive chemical species which rapidly react with organic compounds in the sludge. The main intermediates in this process are:

$$H_2O \rightarrow [2.7]OH^* + [2.6]e^-_{aq} + [0.6]H^* + [2.6]H_3O^+ + [0.7]H_2O_2 + [0.45]H_2 \tag{1}$$

The effect of electron beam irradiation on the microbial reduction in sludge is accomplished by a two-fold effect of the irradiation on the sludge [22]. The first one is the direct effect of radiation on the microorganisms disrupting the structural integrity of DNA molecule and eventually killing them. The other one is the indirect effect caused by the radiolysis of water described above. The chemical active species which are produced in water because of the irradiation will cause oxidative damage on nucleic acids, proteins, and lipids in microorganisms leading to their death. Typically, at a given electron beam radiation dose, the killing of microorganisms occurs at a constant rate over time. Similarly, the effectiveness of the killing of the microbial populations including that of bacteria increases with the radiation dose. The effectiveness of electron beam radiation on the killing of microorganisms is better represented by the so called D_{10} value which is the dose of radiation required to kill 90% of the population of microorganisms present in the sludge sample. Because of the difference in structural complexity of microorganisms, D_{10} values may vary more or less significantly from one microorganism to another. Thus, for *Ascaris ova,* that D_{10} value was determined to be in the order of 0.39 kGy [23]. For other microbial contaminants of sludge including *S. typhimurium* and *E. coli*, these D_{10} values were determined to be 0.3 and 0.34 kGy respectively [4]. Recent studies performed by Engohang-Ndong and his research collaborators have shown that the electron beam technology could be used at industrial scale to eliminate potential microbial pathogens from sewage sludge [9]. The US-based research team showed that a dose of 25.7 kGy was enough to eliminate *Ascaris ova* to a level that is not detectable to available technique used to count the helminth eggs in sewage sludge including the use of Sedgwick Rafter cells to count the detectable ascaris ova. Thus, at that electron beam radiation dose, it was possible to achieve a class A sludge. According to the US Environmental Protection Agency (EPA) standards, to be considered class A biosolids, the sewage sludge must contain less than one *Ascaris* ovum per four gram of sludge dry weight [21]. The dose needed to eliminate fecal coliforms to the norm set by the US EPA to convert sewage sludge to class A biosolids is much lower. The experimental dose determined by Engohang-Ndong and collaborators is 6.7 kGy. In other words, when risks of contamination of sewage sludge by helminth eggs is reduced such as in heavily urbanized areas, the doses needed to convert sewage sludge to class A biosolids that in turn could be used to enrich the soil for landscaping and agricultural purposes is very beneficial and requires low consumption of energy.

6. Economic aspects

An important aspect in the implementation of a new technology such as an electron accelerator in a wastewater treatment plant is to anticipate its impact on the operation costs of the facility and on the environment. With respect to the irradiation of sewage sludge several authors have addressed this issue. A team in Florida reported a cost of $2.50 per 1000 gallons of sludge for a 1.5 MeV electron irradiation facility running at 160 gallons per minute [24], while another group compared gamma and electron beam irradiations for a sample of activated sludge and obtained treatment costs of $\$4.20/m^3$ for gamma irradiation and $\$2.10/m^3$ for electron beam irradiation, which are lower compared to $4.85–$5.19 when using conventional technology at the Central District Wastewater Treatment Facility in Miami Dade County [25]. Furthermore, a comparison was made between irradiation at a dose 6 kGy and incineration of sludge samples and showed a cost of $\$60.87/m^3$ for this latter compared to $\$3.12/m^3$ when using gamma radiation. Similar results have been obtained by Engohang-Ndong et al. using a 3 MeV electron accelerator. In this case these authors reported that the cost to irradiate one cubic meter of sludge to a dose of 25.7 kGy will be $1.26 [9]. That comparative cost analysis tends to show that electron beam irradiation of sludge consumes less energy than other technologies. Furthermore, electron beam irradiation requires processing times, and very importantly more environmentally friendly technology compared to other technologies such as gamma radiation and incineration.

7. Conclusions

While the use of irradiation to control microbial populations in foodstuffs, on medical devices, and on the environment is relatively old, the use of electron beam irradiation to eliminate potential human and animal pathogen from sewage sludge is relatively new. At the present times electron beam irradiation is a mature technology and is capable to provide the required power to irradiate sludge at the flow rates used in a sewage sludge decontamination municipal plant. The quality control techniques are readily available to provide measurements of the dose needed to decontaminate the sludge and the technology has proved its capability to eliminate bacterial populations and ascaris ova to levels that are considered safe for humans and animals in landscape, agricultural, and landfill applications. Analysis of the costs involved in using electron beam technology to decontaminate sewage sludge showed that they are competitive with usual costs involving mechanical and chemical means to treat the sewage sludge. Application of this technology opens new possibilities for major agglomerations worldwide to safely repurpose municipal sewage sludges.

Author details

Roberto M. Uribe[1*] and Jean Engohang-Ndong[2]

1 Emeritus Faculty, Kent State University, Kent OH, USA

2 Department of Biological Sciences, Kent State University at Tuscarawas, New Philadelphia, OH, USA

*Address all correspondence to: ruribe@kent.edu

References

[1] Lessel T, Suess A. Ten year experience in operation of sewage treatment plant using gamma irradiation. Radiation Physics and Chemistry. 1984;**24**:3-11

[2] Sivinski J, Ahlstrom S. Summary of Cesium-137 sludge irradiation activities in the United States. Radiation Physics and Chemistry Special Issue on Irradiation Technology for Environmental Conservation. 1984. p. 14

[3] Trump JG. Energized Electrons Tackle Municipal Sludge. American Scientist. 1981;**69**:276

[4] Borrely SI, Cruz AC, Delmaestro NL, Sampa MHO, Somessari ES. Radiation processing of sewage sludge: A review. Progress in Nuclear Energy. 1998; **33**:3-21

[5] Farooq S, Kurucz CN, Waite TD, Cooper WJ. Disinfection of wastewater: High-energy electron vs gamma irradiation. Water Research. 1993; **27**:1177-1184

[6] Chmielewski AG, Zimek Z, Bryl-Sandelewska T, Kosmal W, Kalisz L, Kaźmierczuk M. Disinfection of municipal sewage sludges in installation equipped with electron accelerator. Radiation Physics and Chemistry. 1995;**46**(4-6):1071-1074

[7] Washino M. Treatment of waste water. Radiation Physics and Chemistry. 1981;**18**:383-388

[8] Hashimoto S, Kawakami W. Application of process control techniques to radiation treatment of waste water. Industrial and Engineering Chemistry Process Design and Development. 1979;**18**:269-274

[9] Jean E-N, Uribe RM, Roger G, Mahinda G, Nickelsen MG, Philip L. Effect of electron beam irradiation on bacterial and *Ascaris* ova loads and volatile organic compounds in municipal sewage sludge. Radiation Physics and Chemistry. 2015;**112**:6-12. DOI: 10.1016/j.radphyschem.2015.02.013

[10] ASTM 51507 Standard on Alanine-EPR Dosimetry System. Annual Book of ASTM Standards. Available from: www.astm.org [Accessed: 2021-12-02]

[11] ASTM 51956 Standard on Thermoluminescence-Dosimetry (TLD). Annual Book of ASTM Standards, Available from: www.astm.org [Accessed: 2021-12-02]

[12] Uribe RM, de la Piedad A, Adem E, Reyes JL. Dosimetry in a pilot plant for bulk disinfestation of grain by electron irradiation. I.-Lithium fluoride in powder. Revista Mexicana de Física. 1980;**26**:421-427

[13] ASTM 51649 Standard Practice for Dosimetry in an Electron Beam Facility for Radiation Processing at Energies Between 300 keV and 25 MeV. Annual Book of ASTM Standards. Available from: www.astm.org [Accessed: 2021-12-02]

[14] ASTM 51650 Standard Practice on Cellulose Triacetate Dosimetry System. Annual Book of Standards. Available from: www.astm.org [Accessed: 2021-12-02]

[15] ASTM 51275 Standard Practice for Use of Radiochromic Dye Film Dosimetry System. Annual Book of ASTM. Available from: www.astm.org [Accessed: 2021-12-02]

[16] Reinthaler FF, Galler H, Feierl G, Haas D, Leitner E, Mascher F, et al. Resistance patterns of Escherichia coli isolated from sewage sludge in comparison with those isolated from human patients in 2000 and 2009. Journal of Water and Health. 2013;**11**:13-20

[17] Young S, Juhl A, O'Mullan GD. Antibiotic-resistant bacteria in the Hudson River Estuary linked to wet weather sewage contamination. Journal of Water and Health. 2013;**11**:297-310

[18] Zhou BW, Shin SG, Hwang K, Ahn JH, Hwang S. Effect of microwave irradiation on cellular disintegration of Gram positive and negative cells. Applied Microbiology and Biotechnology. 2010;**87**:765-770

[19] Kotowska U, Zalikowski M, Isidorov VA. HS-SPME/GC-MS analysis of volatile and semi-volatile organic compounds emitted from municipal sewage sludge. Environmental Monitoring and Assessment. 2012; **184**:2893-2907

[20] Zheng Z, Kazumi J, Waite TD. Irradiation effects on suspended solids in sludge. Radiation Physics and Chemistry. 2001;**61**:709-710

[21] EPA U. 40 Cfr Part 503-Standards for the Use of Disposal of Sewage Sludge. Washington D.C.: Environmental Protection Agency; 2007

[22] Wang J, Wang J. Application of radiation technology to sewage sludge processing: A review. Journal of Hazardous Materials. 2007;**143**:2-7

[23] Capizzi S, Chevallier A, Schwartzbrod J. Destruction of *Ascaris* ova by accelerated electron. Radiation Physics and Chemistry. 1999;**56**:591-595

[24] Kurucz CN, Waite TD, Cooper WJ, Nickelsen MG. Empirical models for estimating the destruction of toxic organic compounds utilizing electron beam irradiation at full scale. Radiation Physics and Chemistry. 1995;**45**:805-816

[25] Meeroff D, Waite T, Kazumi J, Kurucz C. Radiation-assisted process enhancement in wastewater treatment. Journal of Environmental Engineering. 2004;**130**:155-166

Section 2

Biological Technology for Sewage Treatment

Chapter 3

Progress in Domestic Wastewater Treatment, Resource Recovery and Energy Generation Using Microbial Fuel Cell

Girum Ayalneh Tiruye

Abstract

Microbial fuel cells (MFC) are emerging as a versatile eco-friendly bioelectrochemical system (BES) that utilizes microorganisms as biocatalysts to simultaneously convert chemical energy in the chemical bond of organic and inorganic substrates into bioelectricity and treat wastewater. The performance of MFC depends on the electroactive microorganisms, popularly known as exoelectrogens, the loading rate of organic substrate, pH, MFC configurations, hydraulic retention time, and temperature. In most cases, the performance of MFC can be evaluated by measuring chemical oxygen demand (COD) removal efficiency, Coulombic efficiency and MFC power density output. To date, the most common MFC's reactor designs are single-chamber MFC, double-chambers MFC, and stacked-MFC configurations. Generally, considerable developments in MFC systems for waste treatment, renewable energy generation and resource recovery have been made in the last two decades, despite critical challenges of capital cost investment, and low efficiency for large scale applications are impeding MFC from commercialization. This mini-review chapter provides a comprehensive assessment of principles and configurations of MFC, treatment of domestic wastewater, energy generation, and resource recovery by MFC and challenges of MFC. I believe the information provided in this chapter will enlighten the current and future prospects of versatile applications of MFC during domestic wastewater treatment.

Keywords: microbial fuel cell, MFC configuration, domestic wastewater, energy production, wastewater treatment, resource recovery

1. Introduction

The demand for sustainable resources and clean energy with minimal resource consumption is increasing because of rapid global population expansion, rising industrial development, high levels of environmental issues and energy insecurity. The world is confronting a climate change catastrophe and developing technologies that recycle wastes into value-added products, and renewable energy is a critical first step toward addressing the issues.

Waste flow is unending in today's dynamic world; hence, recycling and repurposing waste as a source of value-added materials and clean energy are the comprehensive and intellectual strategy for the future. This optimistic approach of utilizing wastewater as a source of value-added products and clean energy would save society from energy insecurity and environmental resource depletion from the earth. Domestic or municipal wastewater and industrial wastewater are the two main types of wastewater generating every day. Domestic wastewater contains significant amounts of chemical oxygen demand (COD) with the range from 60 to 111,600 mg/L COD. Industrial wastewater, on the other hand, comprises a variety of nutrients depending on the source of the waste or the industry [1]. Water pollution and a lack of sufficient energy are the two most pressing issues today; however, a new technology known as the microbial fuel cell (MFC) can help address these issues in part. MFC is a versatile technology and can be used for a variety of applications, including Electric power generation, wastewater treatment, recovery of pure materials, removal of organic matters, water softening, bioremediations, dye decolorization and biosensor [2–5].

In this mini-review, the principle of MFC, evaluation of MFC's performance in domestic wastewater treatment, and varieties of MFC configurations are all explained. Moreover, the progress of domestic wastewater treatment, energy generation, and resource recovery simultaneously by MFC are all briefly summarized. In addition, the challenges encountered by MFC during the application of domestic wastewater treatment are thoroughly discussed. Finally, concluding remarks on domestic wastewater treatment by MFC are forwarded.

1.1 Fundamental principles of MFC

MFC is an ecofriendly bioelectrochemical system (BES) that utilizes microorganisms as biocatalysts to convert chemical energy in the chemical bond of organic and inorganic substrates into bioelectricity [6]. The organic and inorganic substrates used in MFC as the main feed to generate bioelectricity are low-grade biomasses like lignocellulose, artificial and real wastewaters which are all discharged to the environment every day as waste [7]. As a result, even while major technological challenges remain for its practical development and large-scale applications, MFC is becoming increasingly favorable in terms of environmental sustainability and alternative green electricity generation [8].

MFC mostly comprises anodic and cathodic chambers, and both are separated by proton exchange membrane (PEM). If one of the anodic or cathodic chambers in BES is triggered by microorganisms and produces electrical energy, the term Microbial Fuel Cell (MFC) is used. Microbial electrolysis (MEC) is employed when the system uses electrical energy to accelerate the electrochemical reactions (see **Figure 1**) [10].

1.2 Evaluation of MFC's performance during domestic wastewater treatment

In most cases, the performance of MFC can be evaluated by measuring three parameters: COD removal efficiency, Coulombic efficiency and MFC power density output.

The COD removal efficiency (η_{COD}) of MFC indicates the total energy produced by microorganisms from the substrates (mostly organic matters). It can be estimated by using the following equation:

$$\eta_{COD} = \frac{\left(COD_{inf} - COD_{eff}\right)}{COD_{inf} * 100\%} \quad (1)$$

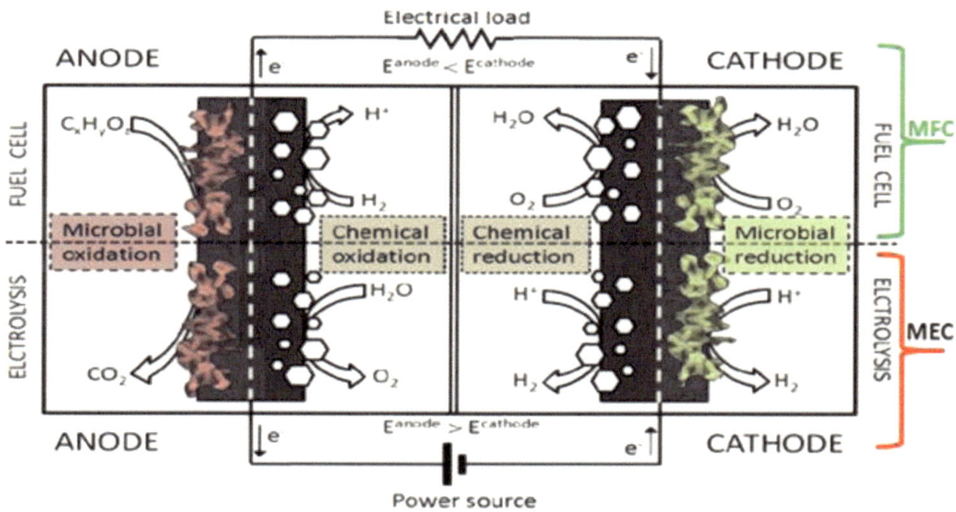

Figure 1.
Schematic overview of the possible combination of microbial and chemical analysis in BESs. Energy is harvested in MFCs if $E_{anode} < E_{cathode}$*, and energy is consumed in MECs if* $E_{anode} > E_{cathode}$ *[9].*

where, COD_{inf} and COD_{eff} are the influent and effluent COD (mg/L) respectively.

All organic matters cannot be converted to useful energy in MFC. This is because the biofilm formed on the MFC's anode or/and cathode chamber needs the energy to grow and maintain itself. As a result, some energy is dissipated as unrecoverable low-grade heat due to overpotentials such as pH gradient across PEM, ionic loss, activation overpotential, concentration overpotential, anode and cathode overpotentials, and ohmic loss due to internal resistance. Therefore, the real potential generated by the closed-circuit MFC is significantly lower than the energy produced from the theoretical open circuit potential. Thus, the performance of MFC is evaluated by calculating the real closed-circuit potential by using the standard potential as follow [8]:

$$U_{output} = E_{cathode} - E_{anode} - \sum \eta_j + I * R_i \qquad (2)$$

where $\sum \eta_j$ is the sum of both activation and concentration overpotentials, I is the current flow, R_i is the internal resistance in the circuit. $E_{cathode}$ and E_{anode} are electrode potentials for cathode and anode, respectively which can be calculated by Nernst equation [6]:

$$E_{cathode} = E^{o}_{cat} - \frac{RT}{nF}\left[\ln\left(\frac{[R]}{[O]}\right)\right], E_{anode} = E^{o}_{an} - \frac{RT}{nF}\left[\ln\left(\frac{[R]}{[O]}\right)\right] \qquad (3)$$

where, E^{o}_{cat} and E^{o}_{an} are standard electrode potentials for cathode and anode, respectively, O is the oxidized species, R is the reduced species and n is the number of electrons that transfers during reaction. R is the universal gas constant (8.314 J/mol K), F is Faraday's constant (9.64853 * 104 C/mol) and T (K) is the absolute temperature.

The second parameter used for evaluating the performance of MFC is done by estimating the Coulombic efficiency. Coulombic efficiency explains the ratio of numbers of electrons transfer vial external resistance, R (ohms), which generates electricity, to the total number of electrons generated from the organic substrate by microorganisms. Therefore, the Coulombic efficiency (CE) of MFC can be calculated as follow [11]:

$$CE = \frac{\left(\int_{t_1}^{t_2} Udt\right)/R}{F * b(\Delta COD)V} * MW \tag{4}$$

where U is the output voltage as a function of time (t), R is external resistance in ohms, b is the number of electrons exchanged per mol of O_2, equal to 4, COD is the removal of chemical oxygen demand, V is the volume of wastewater in litter in the anodic chamber, and MW is the molecular weight of O_2. The last parameter used for measuring the performance of MFC is power density based on electrode projected surface area (PA) or/and power density based on the liquid volume in the anodic or cathodic chamber (PV) and it can be calculated as follow:

$$\mathrm{P} = \mathrm{IU}, \mathrm{PA} = \mathrm{P}/\mathrm{A}, \mathrm{PV} = \mathrm{P}/\mathrm{V} \tag{5}$$

where, A is the surface area of an electrode and V-the liquid volume anolyte in the anodic or catholyte in the cathodic chamber.

2. Domestic wastewater treatment and energy harvesting simultaneously by MFC

Domestic wastewater is any waste that has been used and then discharged into the environment by consumers in any community. It consists of all types of waste materials, such as feces and urine which are added to the water during flushing toilets, personal washing, laundry, food preparation and kitchen cleaning [12].

Domestic wastewater is one of the sources of water, energy and value-added chemicals and nutrients for plant fertilizers, among other things. As a result, it is critical to the process and converts domestic wastewater into renewable energy, value-added products and reduces sludge generation. The aerobic wastewater treatment technique, which is one of the conventional and common treatment methods, faces obstacles such as high operating costs and energy-intensive [1]. Therefore, biological wastewater treatment employing MFC is considered as an alternative technique due to water treatment by removing chemical oxygen demand (COD), recovery of value-added chemicals and electricity generation simultaneously [2, 13].

2.1 Principle of wastewater treatment and energy generation by MFC

The principle of wastewater treatment using MFC is that electrochemical reactions are taking place inside the chamber of MFC and pollutants are removed by exoelectrogenic microorganisms. As the result of these reactions, Gibbs free energy (negative free reaction energy) and release energy (electric or electron release) are spontaneously released. The electromotive force (emf), ΔE^0, can be calculated from the standard free energy as follow [2]:

$$\Delta E^0 = -\left[\sum v_i \Delta G^0_{i,products} - \sum v_i \Delta G^0_{i,reactants}\right]/nF - \frac{\Delta G}{nF} \tag{6}$$

where, $\Delta G^0_{i,products}$ and $\Delta G^0_{i,reactants}$ are the negative free energies of formation of products and reactants (J/mol), respectively, n (moles) of stoichiometry factors of the redox reaction, and F—Faraday's constant (96,485 C/mol).

The useful energy that can be extracted from the reaction of thermodynamic is measured by Gibbs free energy of the reaction. If the system of MFC is generating electricity from wastewater, the theoretical cell voltage of MFC (or emf) can be calculated from the difference between anode and cathode potentials as follow:

$$\Delta E^0_{cell} = \Delta E^0_{cathode} - \Delta E^0_{anode} \tag{7}$$

The cell voltage of MFC will be positive if the Gibbs free energy is negative, indicating that electrical energy generation from the reaction in MFC is spontaneous. For instance, if the wastewater with high content of acetate is used as the organic substrate in MFC, and assume that the concentration of acetate ($[CH_3COO^-]$ = $[HCO_3^-]$ = 10 mM, pH = 7 at 298 K, pO_2 = 0.2 bar), with oxygen reduction, the combined redox reaction will be as follow [14]:

$$\text{Ande} : CH_3OO^- + 4H_2O \rightarrow 2HCO_3^- + 9H^+ + 8e^- \left(E^0 = -0.289 \text{ V vs.SHE}\right) \tag{8}$$

$$\text{Cathode} : 2O_2 + 8H^+ \rightarrow \left(E^0 = -0.85 \text{ V vs.SHE}\right) \tag{9}$$

$$\text{Total} : CH_3OO^- + 2O_2 \rightarrow 2HCO_3^- + H^+ \left(\Delta G = -847.6\frac{\text{kJ}}{\text{mol}}; \text{emf} = 1094 \text{ V}\right) \tag{10}$$

The effective mechanisms and treatment of domestic wastewater by MFC are mainly influenced by several factors such as the configuration of MFC, pH of the electrolyte, the temperature of the electrolyte, electrodes configuration, substrate concentrations, biofilm formation, hydraulic retention time and types of microorganisms [1, 15–18]. Some of the parameters and their effect on the operational performance of MFC are briefly summarized in **Table 1**.

From the different operation parameters affecting the performance of MFC during wastewater treatment, only three configurations of MFC are summarized in the following sections.

2.2 Configurations of MFCs for wastewater treatment

It is very important to evaluate the configurations and designs of MFC for domestic wastewater treatment and power generation simultaneously. This is because configurations can alter the reactor volume, the presence or absence of proton exchange membrane or porous spacer, oxygen supply into cathode chamber and electrode spacing. The most common MFC configurations include single-chamber MFC, double chambers MFC, and stacked MFC configurations [15].

2.2.1 Single chamber MFC

Single chamber MFC (SCMFC) contains only one chamber, which anode chamber, with a proton exchange membrane (PEM) or gas diffusion layer (GDL) which separates anode chamber and cathode electrode [19]. In some cases, SCMFC can be designed without a membrane [20]. During operation, the wastewater to be treated and the biocatalysts, which are microorganisms, are added to the SCMFC's anode chamber. For microorganism's survival and to degrade/oxidize organic sources in the water, the environment of MFC should be anaerobic. Anode and cathode of SCMFC are connected by external wire and electrical ions are transferred through it

Parameter	Effect in the performance of MFC
pH	• Increasing the acid level in the anode chamber inhibits the growth of microorganisms. • Increasing the pH level in the cathode chamber decreases the reduction of O_2 potential. • The average optimal pH is between 8 and 11.
Substrate concentration	• A high level of COD enhances the removal of COD in the wastewater and generates higher power density until the optimum point where microorganisms are inhibited. • A high level of COD decreases the average recovery efficiency of NH_4^+-N and PO_4^{3-}-P.
Temperature	• The optimum temperature for the biofilm formation is between 30 and 45°C. • The lower temperature needs a longer start-up time.
Hydraulic retention time	• Energy generation and COD removal are directly proportional to hydraulic retention time.
Resistance	• The lower the external resistance, the higher generation of energy and removal efficiency of COD. • The higher external resistance delays the time for substrate degradation and reduces electron transport.
Aeration	• Increasing the oxygen in the anode chamber inhibits the COD removal. • The higher the oxygen level in the cathode chamber enhances COD removal and power generation.
Configuration of MFC	• Different configurations, single chamber, double chamber, and stacked MFC have different performances in energy generation.

Table 1.
Different parameters affecting the performance of MFC in wastewater treatment [1, 16–18].

from anode to cathode. The air-porous cathode is directly exposed to oxygen from the atmosphere and electrons are transferred from the anode to it via an external wire to complete the circuit. Oxygen is then serves as an electron acceptor and reduces to produce water after reaction with hydrogen ion. The most common configuration of SCMFC is shown in **Figure 2**.

The electrolyte in the anode chamber serves as a separator in SCMFC without membrane. The efficiency of SCMFC without membrane, on the other hand, is lower than that of SCMFC with a membrane. The reason of decreasing efficiency is that degradation of substrates in the anode chamber occurs aerobically as a result of oxygen diffusion into the anode chamber, leaking of the anolyte, and evaporation [21, 22]. Some of these challenges can be prevented or reduced by utilizing polytetrafluoroethylene (PTFE) diffusion layers on the cathode, which improve oxygen diffusion and water loss [21]. Because there is no cost for the membrane, and if the efficiency of SCMFC without a membrane is enhanced, it will be more advantageous economically than SCMFC with membrane. Moreover, since there is no need for aeration in the cathode chamber, SCMFC is more cost-effective [15, 19].

Using glucose as substrate, graphite carbon brush as anode and 30% Pt coated carbon cloth as a cathode [23], the SCMFC configuration can generate power density up to 2400 mW/m^2 over 50 Ω external resistance. Whereas, when SCMFC is employed for domestic wastewater treatment (pH of 7.3–7.6, and chemical oxygen demand (COD) of 200–300 mg L^{-1}), 766 mW/m^2 power density was generated over 1000 Ω external resistance and from 90–95% COD was removed using carbon cloth as the anode, and 0.5 g/cm^2 Pt coated wet-proof carbon cloth with four PTFE layers on the airside as a cathode [24]. In another report, the domestic wastewater

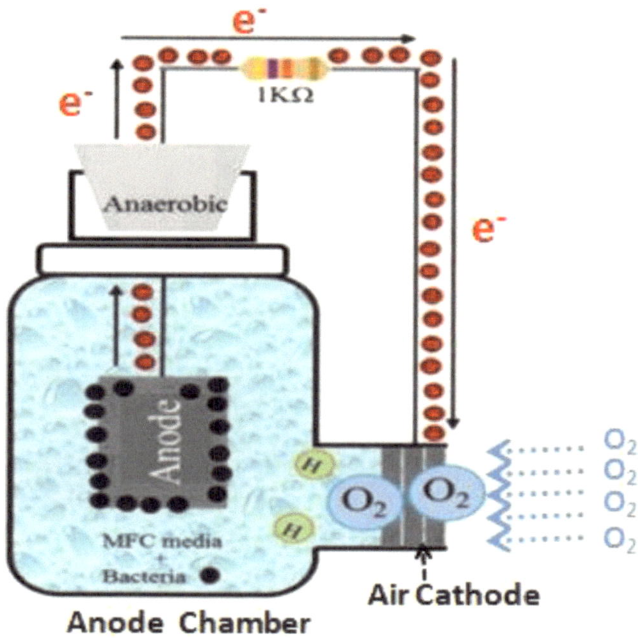

Figure 2.
Configuration of single-chamber MFC (modified from [21]).

(COD = 1010 ± 30 mg L^{-1}) can be modified with acetate and treated by using SCMFC with a single (two-sided) cathode module with a specific surface area of 29 m^2 m^{-3}, two brush anode module and a wire spacer as a separator. The results reveal a higher power density of 1100 mW/m^2 with an average COD removal of 57% [25] than the aforementioned system.

2.2.2 Double chamber MFC

Double chamber MFC (DCMFC) contains anode and cathode chambers which are separated by proton exchange membrane (PEM). PEM allows the transfer of a proton from the anode chamber to cathode chamber while preventing the diffusion of oxygen from cathode chamber to anode chamber (see **Figure 3**). DCMFCs are often used to cleanse wastewater and generation of electricity from the waste simultaneously. Although the cathode and anode chambers are different compartments and separated by PEM, anode and cathode electrodes from each chamber are connected by an external wire through which electrons from the anode are delivered to the cathode [21].

Flat plate MFC [27], bottle MFC [28], miniature MFC [29], and up-flow MFC [30, 31] are examples of DCMFC configurations. The up-flow MFC configurations are thought to be particularly well suited to scaling-up for high larger volume domestic wastewater treatment. However, pumping fluids and recirculation of the waste inside the up-flow MFC systems consume more energy than the system's output, indicating that the primary function of the up-flow MFC is used mainly for wastewater treatment than energy generations. Moreover, another drawback of the DCMFCs configuration is that the catholyte in the cathode chamber must be supplied with new electrolytes regularly or with aeration to provide oxygen to the cathode [21].

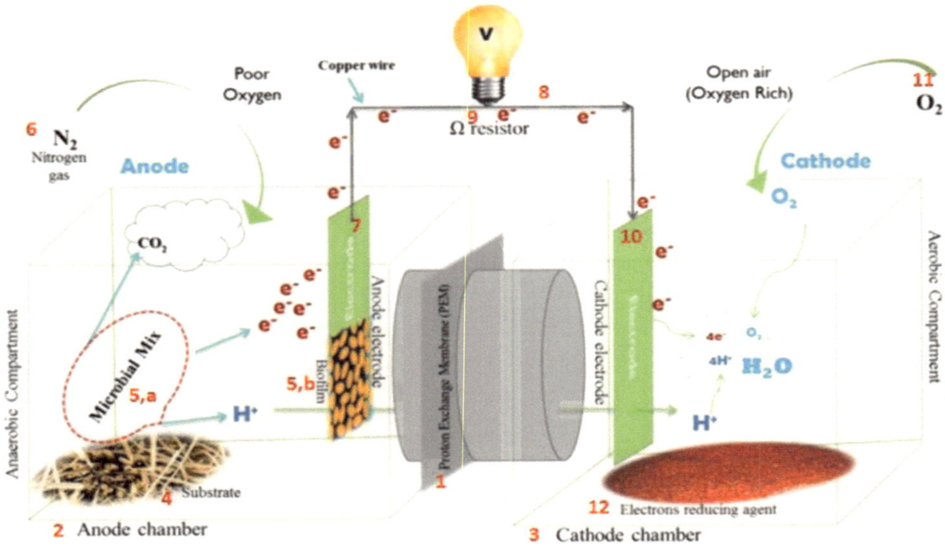

Figure 3.
Major components of double chamber MFC: 1. proton exchange membrane (PEM) selective to H^+ cation and separating the two chambers; 2. anode chamber (anaerobic conditions); 3. cathode chamber under aerobic (open-air) conditions; 4. substrate or biomass for bacteria to feed on; 5. pure or mixed bacterial culture (a) and biofilm (b); 6. nitrogen gas to remove oxygen and maintain the anaerobic condition; 7. anode electrode, on which microorganisms are attached; 8. copper wire for transferring electrons from anode to the cathode; 9. external resistor; 10. cathode electrode; 11. air oxygen; 12. electrons reducing agent [26].

Due to differences in biofilm growth on the anode chamber, treatment of domestic wastewater with DCMFC over different seasons results in variable treatment performances. For example, during the summer season, domestic wastewater treatment generates a higher power density (209 mW/m^2) than during the winter water sample, which generates only 107 mW/m^2. The COD removal efficiency for summer season wastewater was 72%, while there was no significant COD removal for a winter sample of domestic wastewater [32].

2.2.3 Stacked MFC

To scale up the voltage, stacked MFCs combine multiple MFCs in series and parallel. Both connections can be utilized to treat wastewater and generate electricity at the same time. In a series stacked MFC connection, the higher power density and current density can be generated than in a parallel stacked MFC connection [33]. For instance, E.B. Estrada-Arriaga et al. used series and parallel connections of stacked MFCs for domestic wastewater MFC. The results revealed the higher power density (2500 mW/m^2) and current density (500 mA/m^2) in series connection than that of power density (5.8 mW/m^2) and current density (24 mA/m^2) in parallel connection. Moreover, the efficiency of COD removal is also higher in series connection (>80%) than that of the efficiency (>78%) in parallel stacked MFC [34]. The increase of power density and efficiency of COD removal caused by however, in some situations, stacking MFC in series connection faced a challenge due generating negative voltage (voltage reversal), indicating that the system's complex bioreaction resulted in a loss of microbial activity. The challenge can be partially solved by installing diodes in the stacked MFC in order to consume unbalanced electrons and reduce the variation of electrode potentials. On the other hand, for power and wastewater generation, pluggable stacked MFC designs are preferable due to their benefit of off-line capacity for repairs (**Figure 4**) [15].

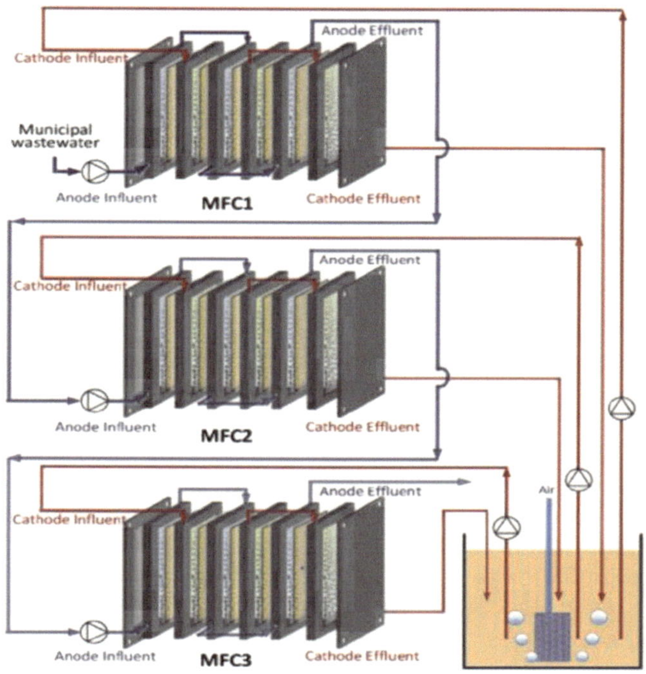

Figure 4.
Water flow connection for treating the municipal wastewater with high COD concentration using stacked MFC. Every three MFC modules were assembled to achieve step-wise COD removal [33].

The largest volume (1000 L) of modularized stacked MFC was assembled and operated for more than a year for municipal wastewater treatment with lower (average 80 mg L^{-1}) and higher initial COD concentrations (average 250 mg L^{-1}). The result showed that the COD removal efficiency was 74% and 70–80% for lower and higher initial COD treatment, respectively. Moreover, the power density was varied in the range from 7 to 60 W m^{-3} (0.42–0.64 W m^{-2}) [33].

3. Resource Recovery from domestic wastewater by MFC

Currently, there is an urgency of closing the cycle of resources for more sustainable development and resource efficiency of current domestic wastewater treatment practices. The effluent of domestic wastewater contains high contents of fat, food residues, detergents, feces, and pharmaceuticals. Thus, in the form of chemicals and value-added compounds, domestic wastewater contains different chemicals presented in **Table 2** [35]. If domestic wastewater is discharged untreated into the water body, the nutrients found in it will be the main potential of causing eutrophication and hypoxia. Therefore, MFCs are useful technologies to recover nutrients from domestic wastewater before its discharge to the environment.

Recovery of these valuable resources presents in domestic wastewater using MFC increases its economic viability. Especially, nitrogen and phosphorous are highly important in the agricultural process due to their use as fertilizers [1].

Nitrogen in domestic wastewater can be removed or recovered using MFC in the form of NH_4^+-N by the processes of ammonification, nitrification and denitrification. Microorganisms in the anode chamber of MFC are the main agents to remove/recover NH_4^+-N and then transport it to the cathode chamber by diffusion and migration. For the diffusion of ammonium, the concentration gradient is the main driver, while

Value-added chemical compounds	Amounts of chemicals
Carbon-rich organic matter (carbonaceous chemical oxygen demand, or COD)	300–600 g
Nitrogen (from ammonium and organic compounds)	40–60 g
Phosphorus (from phosphates and organic compounds)	5–20 g
Sulfur (mainly sulfate) and other traces of heavy metal ions	10–20 g

Table 2.
Different resources from domestic wastewater [35].

electricity is the main parameter in migration. In this case, the content of ammonium in the catholyte is comparative to electricity production by MFC [36–38].

Similarly, phosphorus from domestic wastewater can be precipitated in the form of struvite ($NH_4MgPO_4{\cdot}6H_2O$) on the surface of the cathode [39]. Struvite is a slow-release fertilizer and has many commercial values if it is recovered efficiently including reuse as fertilizers, substituting the demand for phosphorus rock and reducing eutrophication in the water body [40, 41]. The content of phosphorus in struvite is found in the range of 13% and 14% by weight. Since phosphorous is involved in the redox reaction, it can be recovered as struvite according to the following equation.

$$Mg^{+2} + NH_4{}^{+} + PO_4{}^{3-} + 6H_2O \rightarrow NH_4MgPO_4 \cdot 6H_2O\downarrow \qquad (11)$$

The precipitation of struvite occurs near or on the surface of the cathode electrode. This is because the solubility of struvite is decreased at higher pH and in MFC, the pH is higher at the cathode surface due to the reaction producing more OH^- ions as shown in Eq. (12) [17].

$$2H_2O + O_2 + 4e^- \rightarrow 4OH^- \qquad (12)$$

Generally, it is estimated that the availability of total phosphorous in the sewage could supply 15–20% of the world's demand for phosphorous, provided that recovered fully [17].

Therefore, there is now a recommendation of shifting from pollutant removal to resource recovery from domestic wastewater which is now considered as a source rather than a waste [42]. With this regard, MFC is the key technology for resource recovery and energy generation simultaneously from domestic wastewater. For instance, ammonia was recovered with a recovery rate of 3.29 gN d^{-1} m^{-2} and with simultaneous surplus energy generation of 3.46 kJ gN^{-1} from urine by using MFC technology [4].

The loading rate of organic substrates in wastewater significantly affects the recovery/removal of nutrients and electric generation by MFC. This is because, the content of organic substrates in the waste influences the metabolism activities of microorganisms, growth of microorganisms, hence biofilm formation on the surface of electrodes, and the ability of substrate degradation by microorganisms [17]. Yuanyao Y. et al., reported in their experiment that when the loading rate of organic substrate in domestic wastewater is increased, power generation and COD removal are decreased but the recovery efficiency of $PO_4{}^{3-}$-P and $NH_4{}^{+}$-N by MFC are increased. For instance, the maximum power density and COD removal efficiency by MFC at 435 and 870 mg COD/L day loading rate of the organic substrate is 253.84 mW/m^2 with 90% COD removal efficiency and 71.66 mW/m^2 with 70%, COD removal, respectively. On the contrary, when the loading rate of the organic substrate was 435 mg COD/L day, the average efficiency in removing NH_4^+-N and PO_4^{3-}-P in the anode effluent was 14% and 12.43%. When the loading rate of organic substrate was increased to 870 mg COD/L day, the average efficiency in removal of PO_4^{3-}-P and NH_4^+-N in domestic wastewater by using MFC were also increased to 71.5% and 75.13%, respectively. However, for the recovery of NH_4^+-N from domestic wastewater using MFC, the average recovery reduced from 85.11% to 24.34% while increasing the loading rate of organic substrates from 435 to 870 mg COD/L day, respectively. Similarly, the recovery rate of PO_4^{3-}-P was decreased from 83.23% to 24.4% while increasing the loading rate of the organic substrate from 435 to 870 mg COD/L day, respectively [16]. These results strictly showed that the loading rate of organic substrate significantly affects the removal of COD, generation of power and nutrient recovery from domestic wastewater using MFC technology.

4. Challenges of MFC for domestic wastewater treatment

The main drawback of MFC technologies for large-scale applications is the capital cost investment, hence making the technology not feasible for large-scale wastewater treatment and other applications. The costs are mainly associated with electrodes materials, membranes, separators, current collectors and the addition of expensive Pt catalysts [1, 19, 43]. The operational cost for the treatment of domestic wastewater by MFC is as high as 30 times that of the conventional activated sludge treatment of domestic wastewater [44]. However, operational costs can be minimized with the power output of MFC that can be reused for domestic wastewater treatment associated with heating, despite power production is not higher compared with the power input. In addition, the process of domestic wastewater treatment by MFC does not produce a high quantity of sludge, indicating less treatment process of sludge before it is discharged to the environment [1, 19].

Another challenge of MFC for domestic wastewater treatment is scaling-up the size and its treatment efficiency. Scaling-up is mainly expressed in terms of increasing capacity and size of MFC that facilitate and enhance electron transport efficiency. This challenge can be partially solved by large stacked MFC systems with 250–1000 L [33, 45]. Moreover, the distance between anode and cathode also affects the efficiency of MFC because of an increment of resistance. Minimizing the distance between electrodes can solve the issue partially. The electrode stability during domestic wastewater treatment is a very important factor for the performance of MFC. Thus, developing low-cost, high current density output, and carbon-rich anode materials from waste tires are of high importance for solving the problem [43, 46].

5. Conclusion

Domestic wastewater, if left untreated, can be the main source of pollution in the environment, but it can also be used as a raw material for energy production and nutrient recovery. Domestic wastewater treatment, bioelectricity generation and resource recovery simultaneously by using MFC is an eco-friendly strategy and provides many benefits. These include one, it helps the direct generation of renewable electric power from waste. The power generated can be reused for wastewater treatment processes and thus minimizes costs associated with the energy consumption for heating the process. Second, it treats domestic wastewater by removing COD and releases good quality effluents before discharging it to the environment, thus achieving a low environmental footprint. Third, it assists resource recovery like phosphorus, carbon-rich compounds and nitrogen from domestic wastewater because of the effective combination of biological processes and electrochemical processes in bioelectrochemical systems. Especially, nitrogen and phosphorous are highly important in the agricultural process due to their use as fertilizers.

However, the performance of MFC depends on the electroactive microorganisms, commonly known as exoelectrogens. Moreover, the operating parameters such as loading rate of organic substrate, pH, MFC configurations, hydraulic retention time, and temperature all have an impact on MFC performance during domestic wastewater treatment. In most cases, the performance of MFC can be evaluated by measuring three parameters: COD removal efficiency, Coulombic efficiency and MFC power density output. One of the elements in the success of MFC is its design. To date, the most common MFC reactor designs have been single-chamber MFC, double chambers MFC, and stacked MFC configurations. The size of each type of reactor design greatly varies with some MFCs having sizes of a few square centimeters and others having up to a square meter with volumes ranging from milliliters to thousands of liters, respectively. Considerable developments in MFC systems for waste treatment, renewable energy generation and resource recovery have been made in the last two decades, despite critical challenges of capital cost investment, and low efficiency for large-scale applications are impeding MFC from commercialization. Therefore, improving these technical challenges must pave the way for making economically feasible large-scale MFC. Further research will suggest reasonable design, and size of reactors for the multipurpose MFC to efficiently treat domestic wastewater, generate renewable energy and resource recovery.

Acknowledgements

The author is grateful to the work facilities provided by Addis Ababa University, Ethiopia.

Conflict of interest

The authors declare no conflict of interest.

Author details

Girum Ayalneh Tiruye
Materials Sciences Program/Department of Chemistry, College of Natural and Computational Sciences, Addis Ababa University, Addis Ababa, Ethiopia

*Address all correspondence to: girum.ayalneh@aau.edu.et

References

[1] Munoz-Cupa C, Hu Y, Xu C, Bassi A. An overview of microbial fuel cell usage in wastewater treatment, resource recovery and energy production. Science of the Total Environment. 2021;**754**: 142429. DOI: 10.1016/j.scitotenv. 2020.142429

[2] Gude VG. Wastewater treatment in microbial fuel cells—An overview. Journal of Cleaner Production. 2016;**122**: 287-307. DOI: 10.1016/j.jclepro.2016. 02.022

[3] Habermann W, Pommer EH. Biological fuel cells with sulphide storage capacity. Applied Microbiology Biotechnology. 1991;**281**:128-133. DOI: 10.1007/BF00180650

[4] Kuntke P, Śmiech KM, Bruning H, Zeeman G, Saakes M, Sleutels THJA, et al. Ammonium recovery and energy production from urine by a microbial fuel cell. Water Research. 2012;**46**: 2627-2636. DOI: 10.1016/j.watres.2012. 02.025

[5] Jiang Y, Yang X, Liang P, Liu P, Huang X. Microbial fuel cell sensors for water quality early warning systems: Fundamentals, signal resolution, optimization and future challenges. Renewable and Sustainable Energy Reviews. 2018;**81**:292-305. DOI: 10.1016/j.rser.2017.06.099

[6] Logan BE, Hamelers B, Rozendal R, Schröder U, Keller J, Freguia S, et al. Microbial fuel cells: Methodology and technology. Environmental Science & Technology. 2006;**40**:5181-5192. DOI: 10.1021/es0605016

[7] Pant D, Van Bogaert G, Diels L, Vanbroekhoven K. A review of the substrates used in microbial fuel cells (MFCs) for sustainable energy production. Bioresource Technology. 2010;**101**:1533-1543. DOI: 10.1016/j. biortech.2009.10.017

[8] Zhou M, Wang H, Hassett DJ, Gu T. Recent advances in microbial fuel cells (MFCs) and microbial electrolysis cells (MECs) for wastewater treatment, bioenergy and bioproducts. Journal of Chemical Technology and Biotechnology. 2013;**88**:508-518. DOI: 10.1002/jctb.4004

[9] Clauwaert P, Aelterman P, Pham TH, De Schamphelaire L, Carballa M, Rabaey K, et al. Minimizing losses in bio-electrochemical systems: The road to applications. Applied Microbiology and Biotechnology. 2008;**79**:901-913. DOI: 10.1007/s00253-008-1522-2

[10] Rabaey K, Rodríguez J, Blackall LL, Keller J, Gross P, Batstone D, et al. Microbial ecology meets electrochemistry: Electricity-driven and driving communities. The ISME Journal. 2007;**1**:9-18. DOI: 10.1038/ismej.2007.4

[11] Zhou M, Chi M, Wang H, Jin T. Anode modification by electrochemical oxidation: A new practical method to improve the performance of microbial fuel cells. Biochemical Engineering Journal. 2012;**60**:151-155. DOI: 10.1016/ j.bej.2011.10.014

[12] Mara D. Domestic Wastewater Treatment in Developing Countries. In: Mara D, editor. 1st ed. London: Earthscan; 2003. ISBN 1844070190

[13] Logan BE, Rabaey K. Conversion of wastes into bioelectricity and chemicals by using microbial electrochemical technologies. Science (80-.). 2012;**337**:686-690. DOI: 10.1126/ science.1217412

[14] Rozendal RA, Hamelers HVM, Rabaey K, Keller J, Buisman CJN. Towards practical implementation of bioelectrochemical wastewater treatment. Trends in Biotechnology. 2008;**26**:450-459. DOI: 10.1016/j. tibtech.2008.04.008

[15] Gul H, Raza W, Lee J, Azam M, Ashraf M, Kim KH. Progress in microbial fuel cell technology for wastewater treatment and energy harvesting. Chemosphere. 2021;**281**: 130828. DOI: 10.1016/j.chemosphere. 2021.130828

[16] Ye Y, Ngo HH, Guo W, Chang SW, Nguyen DD, Liu Y, et al. Effect of organic loading rate on the recovery of nutrients and energy in a dual-chamber microbial fuel cell. Bioresource Technology. 2019;**281**:367-373, DOI: 10.1016/j.biortech.2019.02.108

[17] Paucar NE, Sato C. Microbial fuel cell for energy production, nutrient removal and recovery from wastewater: A review. Processes. 2021;**9**:1318. DOI: 10.3390/pr9081318

[18] Verma P, Daverey A, Kumar A, Arunachalam K. Microbial fuel cell—A sustainable approach for simultaneous wastewater treatment and energy recovery. Journal of Water Process Engineering. 2021;**40**:101768. DOI: 10.1016/j.jwpe.2020.101768

[19] Christgen B, Scott K, Dolfing J, Head IM, Curtis TP. An evaluation of the performance and economics of membranes and separators in single chamber microbial fuel cells treating domestic wastewater. PLoS One. 2015; **10**:1-13. DOI: 10.1371/journal. pone.0136108

[20] Tan SM, Ong SA, Ho LN, Wong YS, Thung WE, Teoh TP. The reaction of wastewater treatment and power generation of single chamber microbial fuel cell against substrate concentration and anode distributions. Journal of Environmental Health Science and Engineering. 2020;**18**: 793-807. DOI: 10.1007/s40201-020-00504-w

[21] Flimban SGA, Ismail IMI, Kim T, Oh S-E. Review overview of recent advancements in the microbial fuel cell from fundamentals to applications, Energies. 2019;**12**:1-20

[22] DU Z, LI Q, TONG M, LI S, LI H. Electricity generation using membrane-less microbial fuel cell during wastewater treatment. Chinese Journal of Chemical Engineering. 2008;**16**: 772-777. DOI: 10.1016/S1004-9541(08) 60154-8

[23] Logan B, Cheng S, Watson V, Estadt G. Graphite fiber brush anodes for increased power production in air-cathode microbial fuel cells. Environmental Science & Technology. 2007;**41**:3341-3346. DOI: 10.1021/ es062644y

[24] Cheng S, Liu H, Logan BE. Increased performance of single-chamber microbial fuel cells using an improved cathode structure. Electrochemistry Communications. 2006;**8**:489-494. DOI: 10.1016/j. elecom.2006.01.010

[25] He W, Zhang X, Liu J, Zhu X, Feng Y, Logan BE. Microbial fuel cells with an integrated spacer and separate anode and cathode modules. Environmental Science: Water Research and Technology. 2016;**2**:186-195. DOI: 10.1039/c5ew00223k

[26] Flimban SGA, Hassan SHA, Rahman MM, Oh SE. The effect of Nafion membrane fouling on the power generation of a microbial fuel cell. International Journal of Hydrogen Energy. 2020;**45**:13643-13651. DOI: 10.1016/j.ijhydene.2018.02.097

[27] Min B, Logan BE. Continuous electricity generation from domestic wastewater and organic substrates in a flat plate microbial fuel cell. Environmental Science & Technology. 2004;**38**: 5809-5814. DOI: 10.1021/es0491026

[28] Ki D, Park J, Lee J, Yoo K. Microbial diversity and population dynamics of activated sludge microbial communities

participating in electricity generation in microbial fuel cells. Water Science and Technology. 2008;**58**:2195-2201. DOI: 10.2166/wst.2008.577

[29] Ringeisen BR, Henderson E, Wu PK, Pietron J, Ray R, Little B, et al. High power density from a miniature microbial fuel cell using Shewanella oneidensis DSP10. Environmental Science & Technology. 2006;**40**: 2629-2634. DOI: 10.1021/es052254w

[30] Hashemi J, Samimi A. Steady state electric power generation in up-flow microbial fuel cell using the estimated time span method for bacteria growth domestic wastewater. Biomass and Bioenergy. 2012;**45**:65-76. DOI: 10.1016/j.biombioe.2012.05.011

[31] Dessì P, Chatterjee P, Mills S, Kokko M, Lakaniemi AM, Collins G, et al. Power production and microbial community composition in thermophilic acetate-fed up-flow and flow-through microbial fuel cells. Bioresource Technology. 2019;**294**: 122115. DOI: 10.1016/j.biortech.2019.122115

[32] Ali AEH, Gomaa Ola M, Fathey R, El Kareem HA, Zaid MA. Optimization of double chamber microbial fuel cell for domestic wastewater treatment and electricity production. Ranliao Huaxue Xuebao/Journal of Fuel Chemistry and Technology. 2015;**43**:1092-1099. DOI: 10.1016/s1872-5813(15)30032-3

[33] Liang P, Duan R, Jiang Y, Zhang X, Qiu Y, Huang X. One-year operation of 1000-L modularized microbial fuel cell for municipal wastewater treatment. Water Research. 2018;**141**:1-8. DOI: 10.1016/j.watres.2018.04.066

[34] Estrada-Arriaga EB, Hernández-Romano J, García-Sánchez L, Guillén Garcés RA, Bahena-Bahena EO, Guadarrama-Pérez O, et al. Domestic wastewater treatment and power generation in continuous flow air-cathode stacked microbial fuel cell: Effect of series and parallel configuration. Journal of Environmental Management. 2018;**214**:232-241. DOI: 10.1016/j.jenvman.2018.03.007

[35] Li W-W, Yu H-Q, Rittmann BE. Reuse water pollutants. Nature. 2015; **528**:29-31. DOI: 10.1038/528029a

[36] Ye Y, Ngo HH, Guo W, Chang SW, Nguyen DD, Liu Y, et al. Microbial fuel cell for nutrient recovery and electricity generation from municipal wastewater under different ammonium concentrations. Bioresource Technology. 2019;**292**:121992. DOI: 10.1016/j.biortech.2019.121992

[37] Yakar A, Türe C, Türker OC, Vymazal J, Saz Ç. Impacts of various filtration media on wastewater treatment and bioelectric production in up-flow constructed wetland combined with microbial fuel cell (UCW-MFC). Ecological Engineering. 2018;**117**:120-132. DOI: 10.1016/j.ecoleng.2018.03.016

[38] Yang Z, Pei H, Hou Q, Jiang L, Zhang L, Nie C. Algal biofilm-assisted microbial fuel cell to enhance domestic wastewater treatment: Nutrient, organics removal and bioenergy production. Chemical Engineering Journal. 2018;**332**:277-285. DOI: 10.1016/j.cej.2017.09.096

[39] Jadhav DA, Ghosh Ray S, Ghangrekar MM. Third generation in bio-electrochemical system research—A systematic review on mechanisms for recovery of valuable by-products from wastewater. Renewable and Sustainable Energy Reviews. 2017;**76**:1022-1031. DOI: 10.1016/j.rser.2017.03.096

[40] Güney K, Weidelener A, Krampe J. Phosphorus recovery from digested sewage sludge as MAP by the help of metal ion separation. Water Research. 2008;**42**:4692-4698. DOI: 10.1016/j.watres.2008.08.016

[41] Le Corre K, Valsami-Jones E, Hobbs P, Parsons S. Phosphorous recovery from waste water by struvite crysatllisation: A review. Critical Reviews in Environmental Science and Technology. 2009;**39**:433-477

[42] Kehrein P, Van Loosdrecht M, Osseweijer P, Garfí M, Dewulf J, Posada J. A critical review of resource recovery from municipal wastewater treatment plants-market supply potentials, technologies and bottlenecks. Environmental Science: Water Research and Technology. 2020;**6**:877-910. DOI: 10.1039/c9ew00905a

[43] Do MH, Ngo HH, Guo WS, Liu Y, Chang SW, Nguyen DD, et al. Challenges in the application of microbial fuel cells to wastewater treatment and energy production: A mini review. Science of the Total Environment. 2018;**639**:910-920. DOI: 10.1016/j.scitotenv.2018.05.136

[44] He L, Du P, Chen Y, Lu H, Cheng X, Chang B, et al. Advances in microbial fuel cells for wastewater treatment. Renewable and Sustainable Energy Reviews. 2017;**71**:388-403. DOI: 10.1016/j.rser.2016.12.069

[45] Feng Y, He W, Liu J, Wang X, Qu Y, RN. A horizontal plug flow and stackable pilot microbial fuel cell for municipal wastewater treatment. Bioresource Technology. 2014;**156**:132-138. DOI:10.1016/j.biortech.2013.12.104

[46] Chen W, Feng H, Shen D, Jia Y, Li N, Ying X, et al. Carbon materials derived from waste tires as high-performance anodes in microbial fuel cells. Science of the Total Environment. 2018;**618**:804-809. DOI:10.1016/j.scitotenv.2017.08.201

Chapter 4

Secondary Sludge Biodegradation and Electricity Generation in Biocathode Microbial Fuel Cells

Petia Mijaylova Nacheva, Danilo Gamboa-Santana and Edson B. Estrada-Arriaga

Abstract

The looking for sustainable sewage sludge management technology in the wastewater treatment plants, has brought to light the biocathode microbial fuel cells (bMFCs) which allow simultaneous biological stabilization and direct energy generation, avoiding the production of biogas. In the present study, the performance of bMFCs for the treatment of secondary sludge as anodic substrate was evaluated by analyzing the removal of organic matter, destruction of volatile solids and the generation of electrical energy under different operating conditions and applying two types of cathode chambers. The results indicated that VSS and tCOD removals up to 92% and 87% respectively can be achieved in the anodic chamber generating simultaneously energy. Current and power densities of 1.80 ± 0.09 $A \cdot m^{-3}$ and 0.43 ± 0.02 $W \cdot m^{-3}$ respectively were reached, showing that bMFCs are a reliable alternative to generate electricity during the sewage sludge stabilization process. It was revealed that the pH value and the type of cathodic zone are statistically significant factors that influenced the performance of the bMFCs. The obtained results demonstrated that the electrochemical performance of the bMFCs was better at pH value of 6 in the anodic chamber and when aerobic cathode zone was used.

Keywords: Biocathode, electricity, microbial fuel cell, sludge stabilization

1. Introduction

The wastewater treatment plants generate a lot of sludge and numerous approaches have been proposed for their management, such as anaerobic digestion, dewatering, composting, and landfill treatment [1]. The main processes for organic matter removal (accounting for about 55–60% of total BOD_5, COD, or TOC removal) during wastewater treatment are biodegradation, biotransformation, and sorption to activated sludge in the biological steps (like activated sludge and clarification, anoxic/aerobic/aerobic, aerobic/anoxic/oxic, sequencing batch reactors, and membrane bioreactors) [2, 3]. If considering the initial content of volatile solids in sludge of 100%, about 40–60% of BOD_5, COD, or TOC can be degraded during the anaerobic digestion process if mechanical, thermal, chemical, and biological pretreatments are applied [1, 4]. As it is known, the hydrolysis of complex organic matter (particularly the insoluble organic matter) of sludge into dissolved organic matter is the first and the rate-limiting step of anaerobic sludge digestion [5]. Subsequently, the biodegradable dissolved organic matter fraction can be fermented to volatile fatty acids (VFAs), and they are subsequently converted to biogas by methanogens, while the refractory fraction remains in both, the liquid and solid phase of the anaerobic digestate. The looking for more sustainable sewage sludge management technology in the wastewater treatment plants, has brought to light the biocathode microbial fuel cells (bMFCs) which allow simultaneous biological stabilization and direct energy generation, avoiding the production of biogas. The Microbial Fuel Cell (MFC) is a biochemically catalyzed electrochemical system that converts chemical energy to electrical energy by oxidizing the biodegradable organic matter by means of microorganisms via catalytic reactions [6]. The use of biocathodes can enhance the energy generation, and bMFCs can be applied to convert the organic matter in sewage sludge to electricity under ambient temperature, normal pressure, and neutral pH.

Society is facing an increasing demand for energy and has noticed the urgency of changing the energy structure, which today still relies heavily on fossil fuels. The bioenergy is a renewable resource which provides an efficient way of reducing the global warming impact [7]. MFCs are bioenergy source devices that belong to the field of bio-electrochemical systems, and they are considered a sustainable technology since they allow combining the treatment of low value wastes streams, like the wastewater or the sewage sludge, with a direct conversion of the chemical energy into electrical one through bio-electrochemical reactions using microorganism catalysis [8]. MFCs consist of anode and cathode chambers, which are separated by the proton exchange membranes. The power can be generated through the organic matter anaerobic oxidation, performed by electrogenic bacteria in the anode chamber, and reduction of final electron acceptors in the cathodic one [9]. The electrons are transferred to the anode, and they flow to the cathode via a conductive material having an external resistance; the protons migrate through the membrane, and they are reduced by accepting these electrons through the cathode.

Scaling this technology has been difficult and one of the main limitations has been the cost of the cathode materials incorporating precious metals such as Pt and the unsustainable use of ferricyanide as a catalyst independent of the cathode electrolyte [10]. One of the explorations to eliminate these limitations and improve the cathodic stabilization and power generation, enhancing the economic viability and environmentally sustainability of MFC systems, has been the microbial cathode, which uses electro-trophic bacteria as biocatalysts to accept electrons in the cathode substrate [11]. Moreover, this so-called biocathodes enable the use of alternate electron acceptors that can broaden the utility of MFCs and present potential opportunities for the microbially catalyzed conversion of electrical current into various value-added products [10]. Therefore, bMFCs have attracted a lot of attention and they have been considered as a sustainable way to improve the performance of MFC systems.

For the proper MFC performance, a substrate is required in the anode chamber that provides a source of biodegradable carbon and electrons. Generally, any substrate can be used [12], from simple molecules, such as carbohydrates and proteins, to complex mixtures of organic matter, such as those which can be found in the secondary sludge. For a wastewater treatment plant (WWTP), the main source of energy for the equipment is the electricity and this item represents more than 60% of the plant operating costs [13]. The most widely used wastewater treatment process in Mexico is the conventional activated sludge and their electrical energy consumption is 0.10–1.18 $kWh{\cdot}m^{-3}$ [14]. One of the disadvantages of this process is the generation of large amounts of secondary sludge with high content of organic matter that must be properly treated before their disposal; however, due to the complex sludge composition, their treatment is difficult and expensive [15]. That is why the developing of alternative technologies that simultaneously degrade organic pollutants and generate energy directly has been one of the main topics in this research. The studies related to the use of sewage sludge as substrates in bMFCs are still very scarce [13, 16]. The main objective of the presented study was to evaluate the performance of a bMFCs for electricity generation using secondary sludge as anodic substrate, applying different operating conditions, and testing two types of cathodic chamber, aerobic and anaerobic. They were measured and analyzed the power generation, current densities, and coulombic efficiencies, as well as the organic matter removals and the volatile solid destructions.

2. Methodology

2.1 Experimental setup

Ten cylindrical dual chamber bMFCs, 12 cm in diameter and 13 cm in height, were made of plexiglass. Each reactor was divided into two compartments by a Nafion® proton exchange membrane (Nafion 117#, Sigma-Aldrich, London, UK) with a cross-sectional area of 156 cm^2. In order to increase the porosity of the membrane and improve the electrical efficiency of the cell, a pretreatment was performed following the recommendations presented in the reference [17].

The effective volume of each chamber was 0.679 L. All the reactors had anaerobic chambers, 5 of them had aerobic cathodic chambers and 5 had aerobic ones. In the superior part of each chamber there are two holes, one is for the electrode and the other one is for feeding or for reference electrode introduction, or for pH and temperature monitoring. The second hole is sealed in the anaerobic camaras to prevent oxygen diffusion to the anodic chamber, and it is opened in the aerobic ones. To provide homogenization in the anaerobic chambers, recirculation was introduced using peristaltic pumps (Masterflex). For the recirculation two openings were considered on the side of the anaerobic compartments. The mixing in the anaerobic and abiotic cathodic zones were also performed hydraulically using peristaltic pumps. The aeration of the aerobic chambers was performed by air injection and diffusion in the bottom using porous stone diffusers.

Millrose® carbon fiber brushes with twisted titanium wire were used both as anode and cathode, 5.1 cm in diameter and 7.6 cm long, having a projected surface area of 1.46 m^2.

Electrode and PEM pretreatments were performed according to [17, 18]. Both electrodes were connected using a titanium wire (0.5 mm, purity >99.98%, Alfa Aesar, Heysham, UK).

2.2 Experimental design

The experimental design consisted of a full factorial design (2^3), which is a powerful tool that is used to identify the effect of the independent variables on the responses at different levels. The eight experiments were performed twice, and four additional experiments were added with abiotic cathodes, 2 with aerobic cathodic zone and another 2 with aerobic ones.

The independent variables were kind of the cathodic zone (aerobic and anaerobic), VSS concentration of the treated sludge (8 and 16 g/L), and pH in the anodic zone (5 and 6). The experiments were performed twice, in two phases (10 runs in each one). To evaluate the effects caused by the independent variables, the following parameters were determined as response variables: maximum volumetric power density ($PD_{vol\,max}$), maximum volumetric current density ($CD_{vol\,max}$),

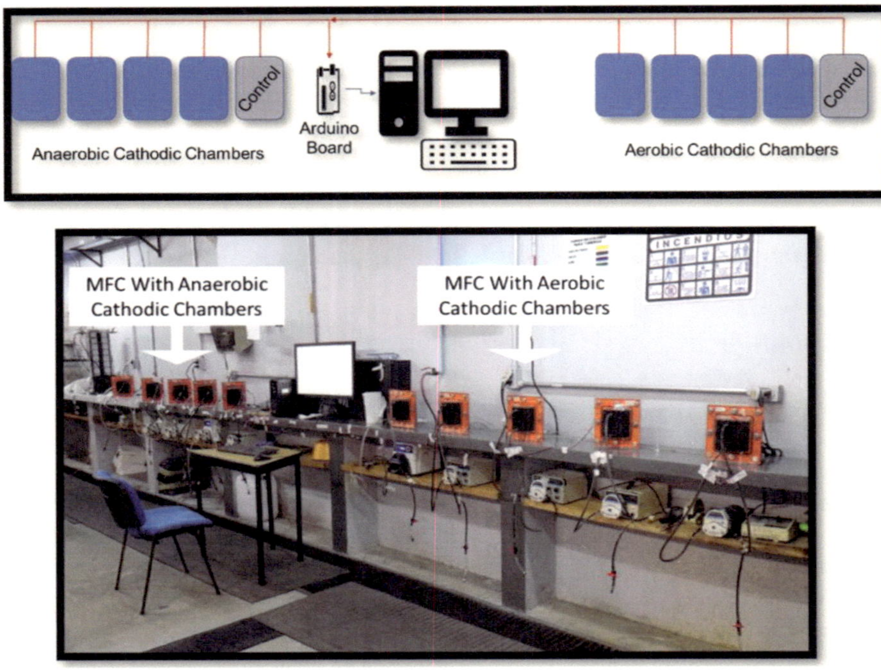

Figure 1.
Schematic diagram of the experimental system.

coulombic efficiency (CE), organic matter removal (R_{COD}), and volatile suspended solids removal (R_{SSV}). The obtained data were analyzed using STHATGRAPHICS Centurion XV ® software. The comparisons of the results were also based on statistical analysis of variance.

The **Figure 1** shows the deployment of the complete experimental system. The left side of the computer shows the anaerobic cathodic zone bMFCs, while the right side of the computer shows the aerobic cathodic zone bMFCs. The purpose of this is to evenly distribute the distance between the computer and the reactors, and to reduce the probability of voltage drops due to the additional resistance that could be caused by the distance of the wiring.

2.3 Inoculation and operating conditions

The experimental bMFCs were operated as sequential batch reactors, allowing the system to adapt to the operating conditions of each cycle (83 days of total operation).

The volume of the catholyte and anolyte were 680 mL each one. The sludge used as substrate for the anode chambers was collected from the secondary settlers of the conventional activated sludge treatment system in one of the wastewater treatment plants located in Mexico City. The sludge was characterized and stored at 4°C until their use as anodic substrate. Two thickened sludge samples were prepared, one with 8 $gVSS{\cdot}L^{-1}$ and another one with 16 $gVSS{\cdot}L^{-1}$. The pH of the sludge was adjusted to 5 and to 6 before the reactor feeding. Milled granular anaerobic sludge from real scale USBR reactor was used as inoculum in the anodic chambers with TS of 155.67 $g{\cdot}L^{-1}$. Almost 5 $gTS{\cdot}L^{-1}$ of inoculum was added to each anodic chamber.

Both catholytes, the aerobic and the anaerobic ones, were formed from a combination of two solutions, so that microorganisms can carry out their metabolic functions: A solution of macronutrients (including the substrate) and a solution of micronutrients (also referred to in this work as a solution of trace elements). The conformations of the first and second solutions are shown in **Tables 1** and 2 [18]. The proportion between both solutions was 8 mL of trace element for each liter of macronutrient solution [19].

The four aerobic cathodic chambers (Aer. C.) were filled with 543 ml of the catholyte and 136 ml of inoculum. Thickened activated sludge from real scale reactor was used as inoculum (SSV of 23231 mg/L, TS of 36875 mg/L). The aerobic cathodic chambers were continuously aerated using porous stone diffusers and aeration system.

A concentration of 733 mg·L^{-1} of sodium nitrite was added to the anaerobic cathodic chambers as final electron acceptor [20]. The four anaerobic cathodic chambers (An.C.) were filled with 648 ml of the catholyte and 32 ml anaerobic inoculum.

The MFC's were operated for three cycles, where each cycle ended when the energy generation dropped below 50 mV. At each cycle change, both the anolyte and the catholyte were exchanged for fresh, new substances. All experiments were performed in duplicate.

The bMFCs were operated at a temperature of 26.9 ± 2.7°C. To start with the experiments, the bMFC's were left operating at open circuit for 165 hours, this allowed the exoelectrogenic microorganisms to adapt to their environment. Subsequently, the electrical circuit was closed by imposing an electrical resistance

Compound	Quantity
NH_4Cl	1.000
K_2HPO_4	1.200
$MgSO_4$	0.500
KCl	0.500
KH_2PO_4	0.140
$Fe_2(SO_4)_3{\cdot}H_2O$	0.010
Yeast extract	0.020

Table 1.
Macronutrientes (in g/L).

Compound	Quantity
$FeSO_4{\cdot}7H_2O$	1000.0
$ZnCl_2$	70.0
$MnCl_2{\cdot}4H_2O$	100.0
H_3BO_3	6.00
$CaCl_2{\cdot}6H_2O$	130.0
$CuCl_2{\cdot}2H_2O$	2.0
$NiCl_2{\cdot}6H_2O$	24.0
$Na_2Mo_4{\cdot}2H_2O$	36.0
$CoCl_2{\cdot}6H_2O$	238.0

Table 2.
Trace elements (in mg/L).

of 100 ohms to all the bMFC's and based on the voltage reading under this condition, we proceeded to calculate the response parameters.

2.4 Analytical methods

The organic matter removals and the volatile solid degradations were calculated based on the obtained results. The sludge stabilization was followed determining the total chemical oxygen demand (TCOD) and volatile suspended solids (VSS). For the determination of TCOD and VSS content, analytical techniques were used according to standard methods [21].

For all the bMFC's, simultaneously the voltage (V) generation was recorded every hour during the whole operation time of each of the 3 cycles, for this basic data acquisition system was designed, programmed, and implemented. Its assembly consisted of a development board based on the ATmega2560 microcontroller, better known as "Arduino Mega 2560" and electronic accessories such as the prototyping board, 22-gauge parallel cable of two soft copper conductors with individual thermoplastic polyvinyl chloride insulation and joined by a track of the same material (commonly known as duplex cable), alligators, digital temperature sensor and jumper wires with male–male terminals. To avoid data loss in the event of a power outage to the laboratory, the computer was connected to a backup power supply capable of supplying power for two hours without interruption.

The current (I) was calculated using Ohm's law and the electrical power (P) with the formula P = V I. The maximum current density (CDmax) and the maximum power density (PDmax) were normalized with the electrode area.

The CE was calculated according to Eq. (1) where Ui is the recorded voltage in volts (V) of the bMFC at time i, in seconds (s), R is the external resistance, in ohms (Ω), F is the Faraday constant (96485. 3365 C $\cdot mol^{-1}\ e^{-}$), b is the number of moles of electrons exchanged per mole of oxygen used in the degradation of organic matter (4 mol $e^{-} \cdot mol^{-1}\ O_2$), ΔS is the removed concentration of COD (mg $O_2 \cdot L^{-1}$), V is the volume of the anolyte, in liters (L), and M is the molecular weight of oxygen (32,000 mg $O_2 . mol^{-1}$) [18, 22, 23].

$$CE(\%) = \frac{\sum_{i=1}^{n} U_i t_i}{RFb\Delta SV_{anol}} M \times 100 \tag{1}$$

Instrument	Manufacturer	Model
Portable multimeter	HACH	HQ40d
Laboratory Low Maintenance Gel Filled pH Electrode	HACH	IntelliCAL PHC101
Laboratory 4-Poles Graphite Conductivity Cell	HACH	IntelliCAL CDC401
Laboratory Spectrophotometer for water analysis	HACH	VIS DR2800
Peristaltic pump	MASTERFLEX L/S economy drive	HV-77916-10
Potentiostat	GAMRY INSTRUMENTS	Interface 1010E

Table 3.
Instruments used in this study.

Cycic Voltammetry analysis was done during the last batch cycle of each bMFC in the voltage range of −0.8 V to +0.8 V at the scan rate of 1 $mV \cdot s^{-1}$. This technique was conducted using a potentiostat with the cathode as the working electrode, the anode as counter electrode and an Ag/AgCl reference electrode, and it was used to evaluate the oxygen reduction reaction catalytic activity of the bMFC with aerobic and anaerobic cathodic zone. The instruments used for analyses, operation and measurements are presented in **Table 3**.

3. Results and discussion

3.1 Performance of the biocathode microbial fuel cells at open circuit

The startup of all the bMFCs was performed at open circuit conditions, with a cycle duration of 165 h. The voltages increased over the time in the bMFCs with aerobic cathodic chambers (Ae.C.Ch.), while a contrary tendency was observed in the bMFCs with anaerobic cathodic chambers (An.C.Ch.). The maximum voltages obtained at different experimental conditions are presented in **Figure 2**. Higher maximum voltages were obtained with aerobic cathodic chambers. For pH of 5, higher voltages were obtained with the lower VSS concentrations, but at pH of 6, the higher VSS concentration allowed obtaining of higher voltages.

The maximum voltages in the bMFCs were lower than those obtained in the MFCs with abiotic cathodic chamber; higher voltage was obtained only in the bMFCs with aerobic cathodic chamber, operated with a VSS concentration of 15 $g \cdot L^{-1}$ and pH 5. The analysis of the average voltages indicated that there was not statistically significant difference between the values obtained in the MFCs with biocathodes and in MFCs with abiotic cathodes.

3.2 Performance of the biocathode microbial fuel cells at closed circuit

The next operating cycles were performed at closed circuit with resistances of 100 Ohms and the obtained results are presented at **Figure 3** and the maximum voltages obtained at each operational condition are illustrated in **Figure 4**.

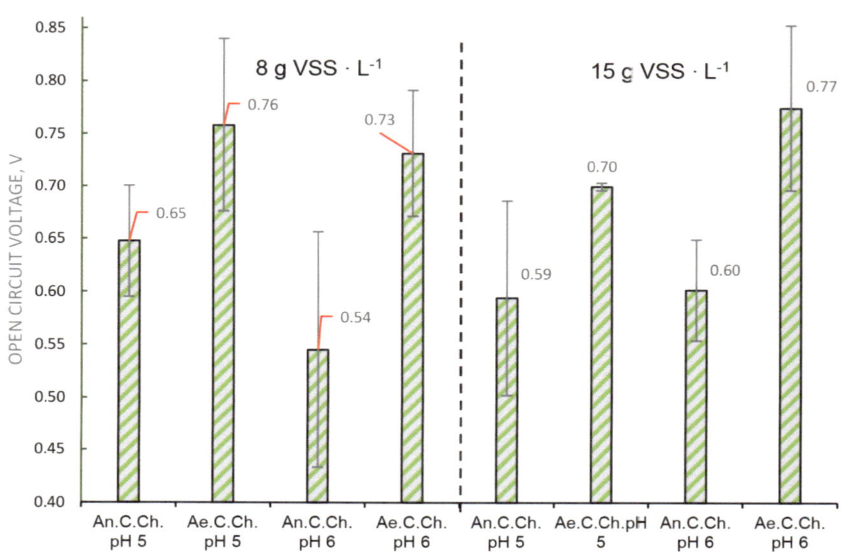

Figure 2.
Maximum voltages reached at open circuit during the first operational cycle.

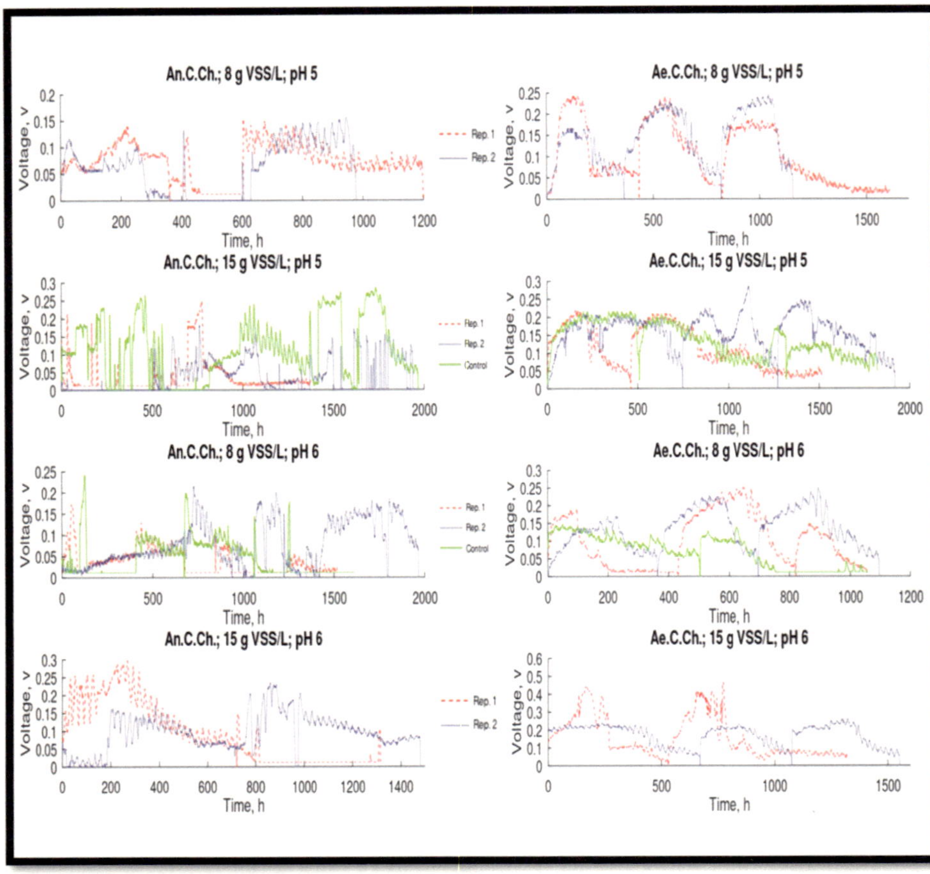

Figure 3.
Obtained voltages in the experimental reactors during the closed-circuit operating cycles.

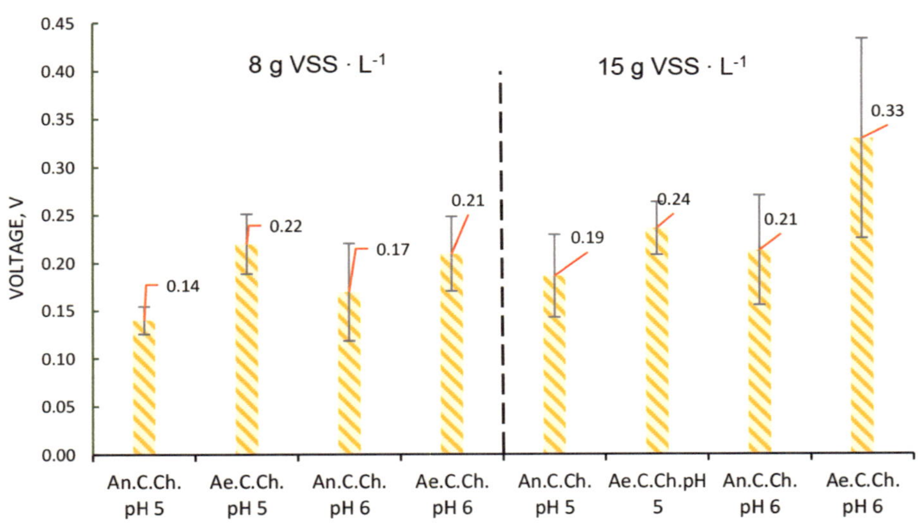

Figure 4.
Maximum voltages obtained during the closed-circuit operating cycles.

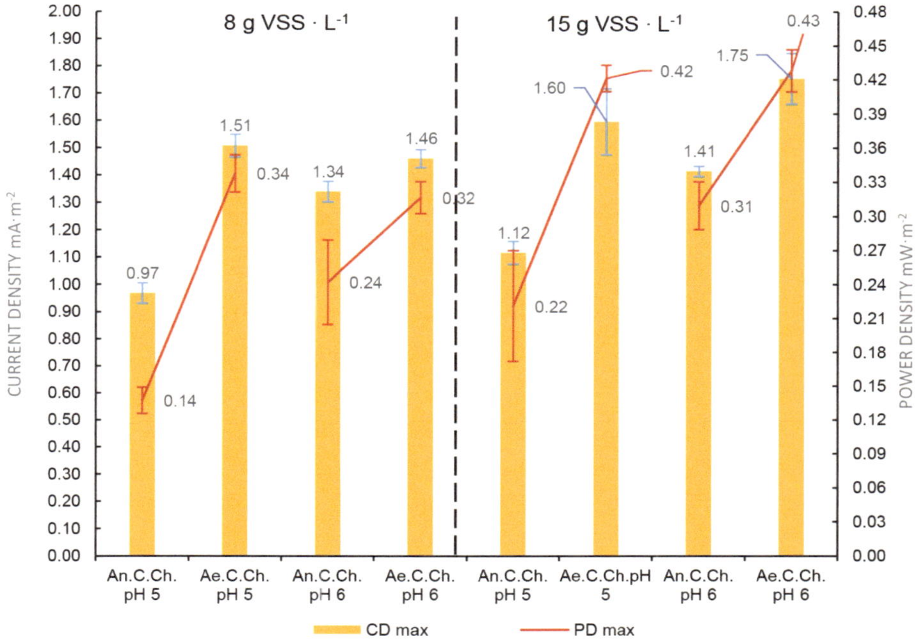

Figure 5.
Maximum volumetric current and power densities achieved during the experiments.

As it can be seen higher voltage were obtained when the oxygen was the final electron acceptor in the cathodic chamber. With respect to the initial concentration of the substrate, the highest VSS concentrations allowed increasing of the obtained voltages. The highest voltages of 0.33 ± 0.03 V were reached with the bMFCs operated with a sludge initial concentration of 15 g VSS·L^{-1} and pH of 6.

To corroborate the statistical difference between the effects of each of the factors on the response variables, volumetric power density (PD) and volumetric current density (CD), regression analysis and ANOVA were carried out. The analyzed data regarding volumetric power density and current density are shown in **Figure 5**. Second order polynomic models were used with determination coefficients (R^2) of 0.925 for PD and 0.922 for CD and adjusted determination coefficient (adj-R^2) of 0.876 for DC and 0.87 for DC, which indicated a good capability of the models to predict the responses within the proposed experimental ranges. There is statistically significant difference between the results obtained with different type of cathodic zones, being the aerobic one that allowed obtaining of higher values for both, PDmax and CDmax, compared with the determined in the reactors with anaerobic cathodic zone (**Figure 6**). The effects of the VSS concentrations and pH values were much lower, but statistically significant, greater results were obtained using sludge with higher VSS concentration (of 15 g·L^{-1}) and higher pH (pH of 6). The lowest effect on PD was the one of pH and on CD was the one of VSS concentration.

A comparison of the electrical power generation in bMFCs and in MFCs with abiotic cathode is presented in **Figure 7**. As it can be observed, higher DCmax and PDmax were obtained in bMFCs with aerobic cathodic chambers operated with 8 g VSS·L^{-1} and pH of 6. However, based on the average values, it was obtained that the difference is not statistically significant for PD (p-value = 0.116059), but it is significant (p-value of 0.0874862) for CD, in favor of the system that used biocathode, applying alpha value of 0.1. Statistical t-tests performed for the case of bMFCs

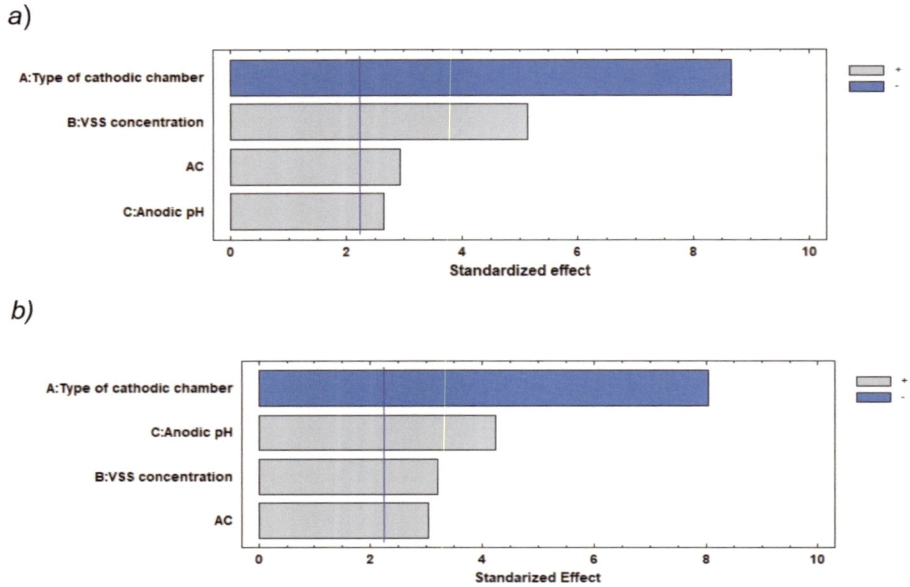

Figure 6.
Standardized Pareto diagrams for PD (a) and for CD (b). For the cathodic zone chamber, the blue color represents the effect of Ae.C.Ch.

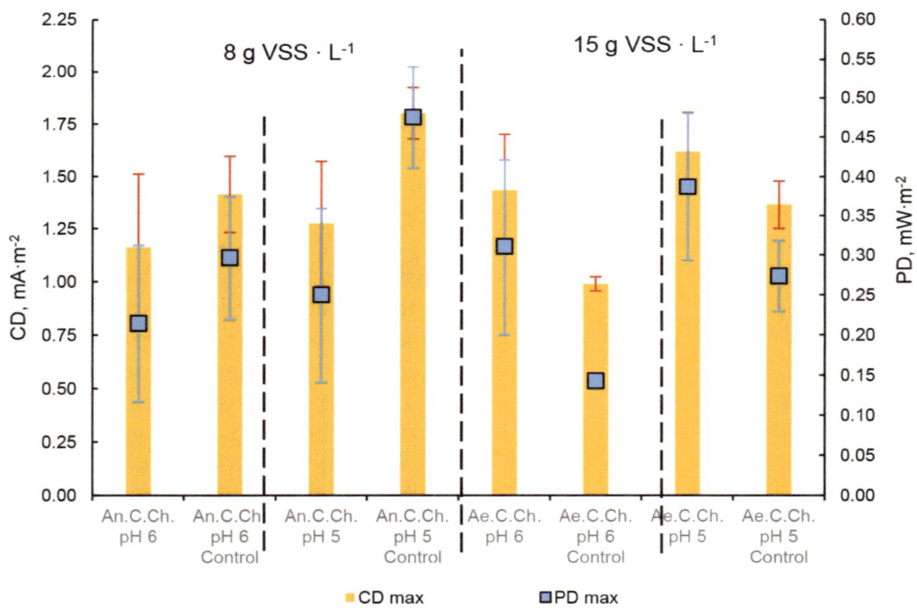

Figure 7.
Comparison of power generation with respect to abiotic cathode controls.

with aerobic cathodic chambers and MFCs with abiotic cathode, operated with 16 g VSS·L^{-1} and pH of 5, showed that there is no statistically significant difference between the results with and without biocathodes for both variables (p-value >0.1).

For bMFCs with anaerobic cathodic chambers operated with 8 g SSV·L^{-1} and pH of 6, the statistical tests indicated that there was not statistically significant difference between the use of abiotic cathode or biocathode (p-value >0.1). However, for bMFCs with anaerobic cathodic chambers operated with 16 g SSV·L^{-1} and pH of 5, the values of both PD and the CD were lower than the obtained in the MFCs with abiotic cathode (p-value of 0.0302892 and 0.0455172 respectively) with a 90% of confidence.

3.3 Organic matter removal and Coloumbic efficiency

The average TCOD and VSS removals determined in the anodically processed sludge, and the coulombic efficiencies obtained using different initial pH and VSS concentrations, and in bMFCs with aerobic and anaerobic chambers, are illustrated in **Figure 8**. As it can be seen the obtained TCOD and VSS removals were higher than 75%, reaching values up to 92%, which indicates that the sludge stabilization process was successful in all the operational conditions.

Based on the obtained results for TCOD removal, the empirical relationship between the response and variables was expressed by a polynomial equation, with determination coefficient (R^2) of 0.997 and adjusted determination coefficient (adj-R^2) of 0.982, which indicated a good capability of the model to predict the response within the proposed experimental ranges. The calculated effect analysis, with a 96% of confidence and a factor of significance (p) of 0.04, indicated as the most significant factor the interaction of pH with the concentration of VSS, both with positive effects.

For the VSS removal, the model had R^2 = 0.999 and adj-R^2 of 0.997, and with a 97% of confidence, the initial VSS concentration had the major effect, followed by the pH. There was not statistically significant effect of the type of the cathodic zone. There was a statistically significant difference between the VSS removals obtained with different initial VSS concentrations (p = 0.03) and with different initial pH values in the anode chamber, being p of 0.04 and 0.03 respectively. There was not statistically significant difference between the results obtained in bMFCs with different type of cathodic zone.

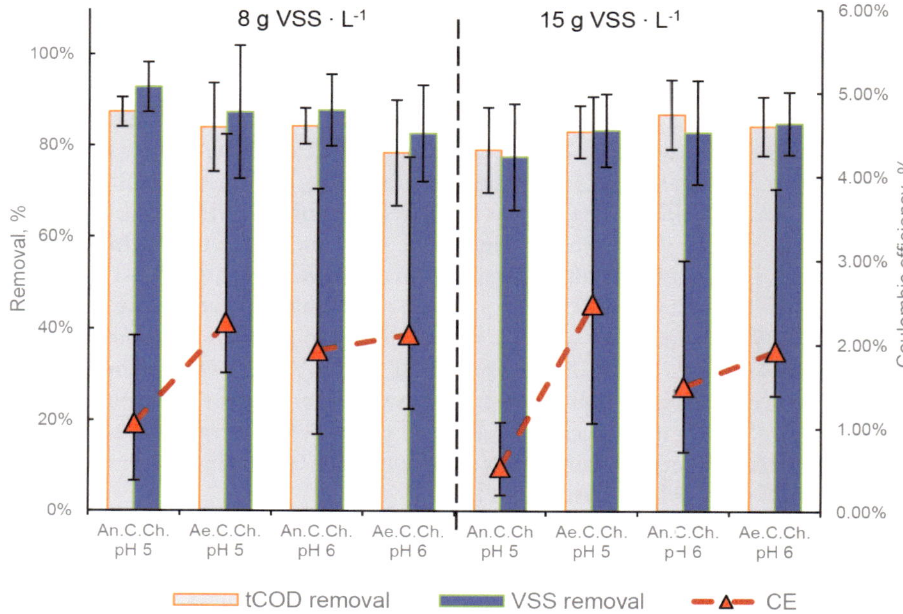

Figure 8.
TCOD and VSS removals, and coulombic efficiencies obtained in the experimental bMFCs.

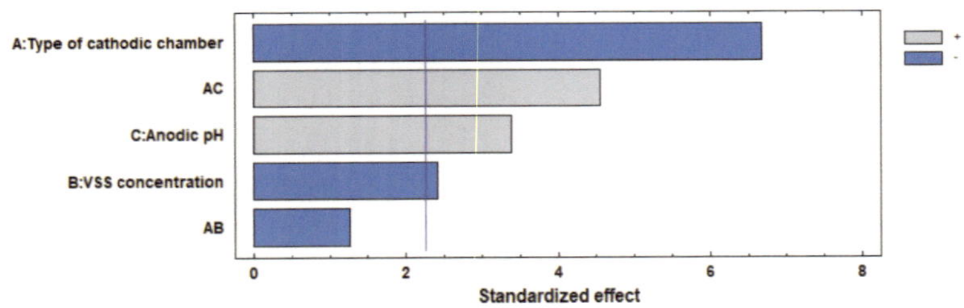

Figure 9.
Standardized Pareto diagrams for CE.

The performed statistical analysis for the Coulombic Efficiency (CE) using a model with R^2 of 0.975 and adj-R^2 of 0.941, indicated that the factors which strongly influenced CE are the type of cathodic zone and the initial pH (**Figure 9**). The use of aerobic cathodic zone and initial pH of 6 allows obtaining of better coulombic efficiencies. The highest coulombic efficiency averages reached in the bMFCs were 2.1–2.3%. These results are lower than those reported by Zhang et al. [18], who reached CE up to 19.4%, although with removals of TCOD up to 40.8%, using a double chamber MFCs with sludge as anodic substrate and biocathodes. On the other hand, the obtaining of small CE values with high COD removals indicates that the electrons released from the organic matter were consumed by processes other than those carried out by electrogenic biofilms, such as fermentative and methanogenic biofilms [24]. This is also consistent with the results obtained by Freguia et al. [25], who indicated that fermentation and methanogenesis are not electrode-dependent reactions, so they could occur with any external resistance as long as the redox potential in the solution is low enough and there are bacteria present that derive more energy from these processes than from electrode-driven oxidation of the substrate. Further research is needed to better understand competitive microbial processes such as exoelectrogenic, biomass growth, fermentative, and methanogenic at the anode to minimize their effects and increase power generation and CE.

3.4 Electrochemical performance of the bMFCs

Polarization curves and power curves generated in the experimental bMFCs are presented in **Figures 10** and **11** respectively. The values of maximum power (Pmax), internal resistance (IR) and open circuit voltages, obtained from the figures are reported in **Table 3**. The graphs indicated that bMFC with aerobic cathodic chamber, initial sludge VSS concentration of 15 g $\cdot L^{-1}$ and pH of 6 was the one who obtained the highest open circuit voltage (553 mV), as well as the highest maximum PD of 0.21 $mW{\cdot}m^{-2}$ at a CD of 0.55 $mA{\cdot}m^{-2}$. The maximum power in the rest of the bMFCs was 0.05–0.15 $mW{\cdot}m^{-2}$ with current densities of 0.25–0.60 $mA{\cdot}m^{-2}$. The maximum CD up to 0.67 $mA{\cdot}m^{-2}$ and PD of 0.14 $mW{\cdot}m^{-2}$ was reached with the lowest applied resistance (46 Ohms) in bMFC with anaerobic cathodic chamber, initial sludge VSS concentration of 15 g $\cdot L^{-1}$ and pH of 5. These values are lower than with the reported for MFCs with similar structural and biotic characteristics which reached 38 mW m^{-2} [26], 43.6 mW m^{-2} [27].

The results of **Table 4** show that for all reactor configurations, both, the internal resistance, and the open circuit voltages (OCV) were higher when the aerobic cathodic chamber was used. For the case of maximum power density, the results do not show a clear pattern that could help relate the configuration to the observed result.

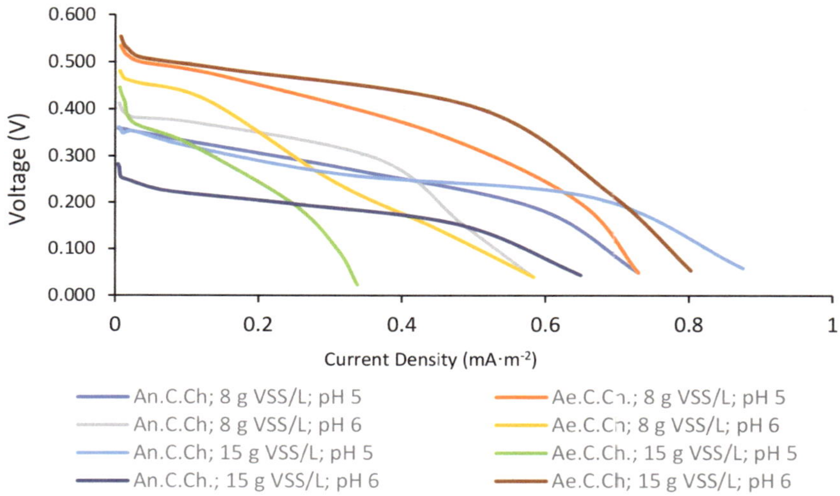

Figure 10.
Polarization curves.

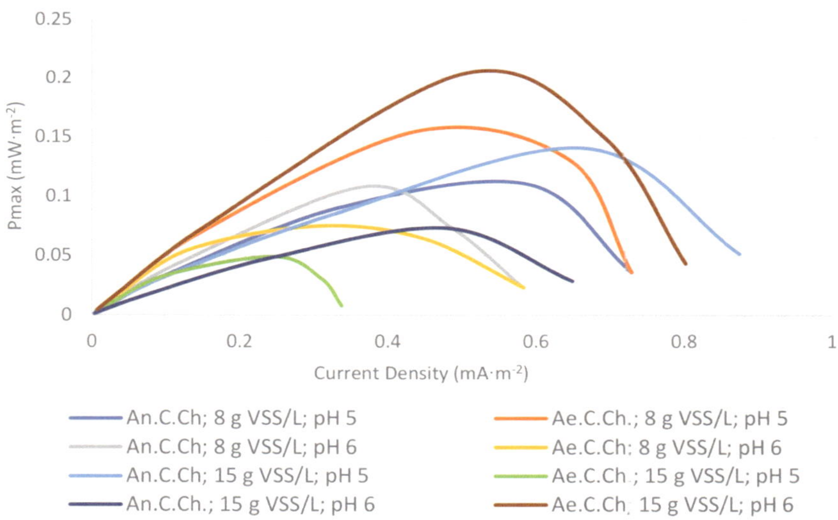

Figure 11.
Power curves.

The voltamperogram generated from all evaluated operating conditions is shown in **Figure 12**. In green, the operating conditions that included Aerobic Cathodic Chamber (Ae.C.Ch.) and in purple those that included Anaerobic Cathodic Chamber (An.C.Ch.). It was also found that at the potentials −800 mV and 800 mV the maximum and maximum current densities of −21.1 μA·cm^{-2} and 7.11 μA·cm^{-2} were achieved (corresponding to −307.9 and 103.8 mA shown in the graph) respectively. These values correspond to the Ae.C.Ch. configuration; 8 g VSS·L^{-1}; pH 6 and agree with the previous analysis for CDmax where the type of cathodic zone and the pH have significant effects. The shape of the graph agrees with what is observed in [28] where the reduction of oxygen shows the fall of the curve on left section. Under anaerobic conditions this does not happen as markedly since nitrogen has a lower oxidation capacity than oxygen.

bMFC	R_{int} (Ohms)	PD_{max} ($mW{\cdot}m^{-2}$)	OCV (mV)
An.C.Ch; 8 g VSS/L; pH 5	223.920	0.110	357.0
Ae.C.Ch; 8 g VSS/L; pH 5	498.200	0.150	534.0
An.C.Ch; 8 g VSS/L; pH 6	207.990	0.110	412.0
Ae.C.Ch; 8 g VSS/L; pH 6	484.430	0.070	480.0
An.C.Ch; 15 g VSS/L; pH 5	97.360	0.140	361.0
Ae.C.Ch; 15 g VSS/L; pH 5	1081.320	0.050	446.0
An.C.Ch; 15 g VSS/L; pH 6	132.410	0.070	282.0
Ae.C.Ch; 15 g VSS/L; pH 6	154.250	0.210	553.0

Table 4.
Internal resistance, OCV and maximum power density.

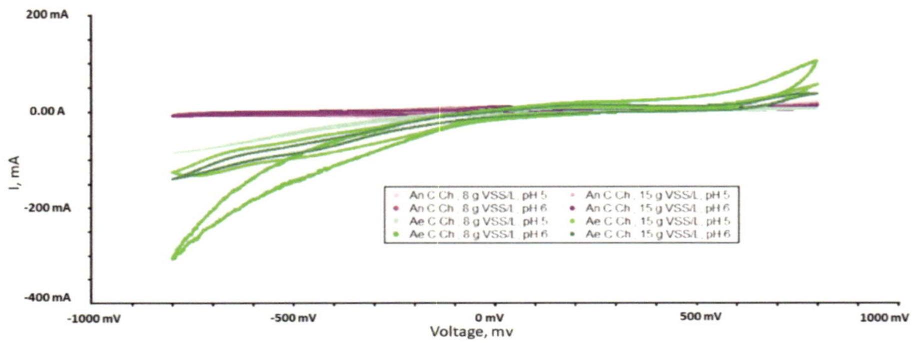

Figure 12.
Cyclic voltammetry obtained in the experimental bMFCs.

4. Conclusions

This study showed that it is possible to stabilize successfully secondary sludge with VSS concentration up to 15 $g{\cdot}L^{-1}$ using microbial fuel cells with biocathodes. The highest organic matter removals reached in the anode chamber, up to 92%, indicates a very good microbial activity in the anodic chambers. The statistical analyses of the obtained results indicated that the kind of the anodic chamber, the variations of VSS concentration in the secondary sludge between 8 and 15 $g{\cdot}L^{-1}$, as well as the variations of pH between 5 an6 do not influenced significatively the organic matter removal.

The simultaneous generation of electricity is possible together with the degradation of organic matter, which contributes to the sustainability of this method. The bio-cathodic microbial fuel cells with aerobic cathodic chambers allows obtaining of higher voltages, current densities, power densities and coulombic efficiencies compared with the microbial fuel cells with anaerobic cathodic chambers (up to 450 mV, 0.43 $mW{\cdot}m^{-2}$, 1.80 $A{\cdot}m^{-2}$ and 4% respectively).

The effects of the VSS concentrations and pH values were much lower than the type of the cathodic chamber, but statistically significant, greater results were obtained using sludge with higher VSS concentration (of 15 $g{\cdot}L^{-1}$) and higher pH (pH of 6). The lowest effect on the power density was the one of pH and on the current density was the one of VSS concentration. The best configuration for operating MCCs varied according to the parameter of interest that is desired as a response variable. For the power density, the best configuration is aerobic cathodic chamber, 15 g $VSS \cdot L^{-1}$ and sludge pH of 5. For the current density the best configuration is aerobic cathodic chamber, 15 g $VSS \cdot L^{-1}$ and pH of 6. For the case of the coulombic efficiency, the best configuration is aerobic cathodic chamber, 8 g $VSS \cdot L^{-1}$ and pH of 6,

The performance comparison of the microbial fuel cells with biotic and abiotic cathodes indicated that there was not statistically significant improvement of the response parameters, and there was even a configuration (anaerobic cathodic chambers, 16 g $SSV \cdot L^{-1}$ and pH of 5) for which better results were obtained with abiotic cathode.

The electrochemical tests confirmed that the configuration with aerobic cathodic chamber, initial sludge VSS concentration of 15 g $\cdot L^{-1}$ and pH of 6 was the one who obtained the highest open circuit voltage (553 mV), as well as the highest maximum power density of 0.21 $mW \cdot m^{-2}$ and current density of 0.55 $mA \cdot m^{-2}$. The internal resistance and the open circuit voltages were higher when the aerobic cathodic chambers were used.

Acknowledgements

The authors would like to thank the financial support of this study through the project A1-S-26278 obtained from the SECTORAL RESEARCH FUND FOR EDUCATION of SEP-CONACYT in Mexico, as well as to Mexican Institute of Water Technology for the infrastructure provided to perform this research.

Author details

Petia Mijaylova Nacheva[1]*, Danilo Gamboa-Santana[2] and Edson B. Estrada-Arriaga[1]

1 Mexican Institute of Water Technology (IMTA), Jiutepec, Morelos, Mexico

2 Mexican National Autonomous University (MrSc Environmental Engineering, Campus IMTA), Mexico

*Address all correspondence to: petiam@tlaloc.imta

References

[1] Zhen G, Lu X, Kato H, Zhao Y, Li YY. Overview of pretreatment strategies for enhancing sewage sludge disintegration and subsequent anaerobic digestion: Current advances, full-scale application and future perspectives. Renew. Sust. Energ. Rev. 2017; 69: 559-577. DOI: 10.1016/j.rser.2016.11.187

[2] Carstea EM, Bridgeman J, Baker A, Reynolds DM. Fluorescence spectroscopy for wastewater monitoring: A review. Water Res. 2016; 95: 205-219.

[3] Xue S, Jin W, Zhang Z, Liu H. 2017. Reductions of dissolved organic matter and disinfection by-product precursors in full-scale wastewater treatment plants in winter. Chemosphere. 2017; 179: 395-404. DOI: 10.1016/j.chemosphere.2017.02.106

[4] Maspolim Y, Zhou Y, Guo C, Xiao K, Ng WJ. Comparison of single-stage and two-phase anaerobic sludge digestion systems –Performance and microbial community dynamics. Chemosphere. 2015; 140: 54-62. DOI: 10.1016/j.chemosphere.2014.07.028

[5] Alvarado A, West S, Abbt-Braun G, Horn H. Hydrolysis of particulate organic matter from municipal wastewater under aerobic treatment. Chemosphere. 2021; 263, 128329. DOI: 10.1016/j.chemosphere. 2020. 128329

[6] Mohan SV, Velvizhi G, Modestra JA, Srikanth S. Microbial fuel cell: critical factors regulating bio-catalyzed electrochemical process and recent advancements. Renewable and Sustainable Energy Reviews. 2014. 40: 779-797. DOI: 10.1016/j.rser.2014.07.109

[7] Rai M, Ingle AP. Sustainable bioenergy: advances and impacts. 1st ed. Elsevier; 2019. 416 p. eBook ISBN: 9780128176559. DOI:

[8] Harnisch F, Schröder U. From MFC to MXC: Chemical and biological cathodes and their potential for microbial bioelectrochemical systems. Chemical Society Reviews. 2010; 39(11): 4433-4448. DOI: 10.1039/c003068f

[9] Du Z, Li H, Gu T. A state of the art review on microbial fuel cells: A promising technology for wastewater treatment and bioenergy. Biotechnology advances. 2007; 25: 464-482. DOI: 10.1016/j.biotechadv.2007.05.004

[10] Huang L, Regan JM, Quan X. Electron transfer mechanisms, new applications, and performance of biocathode microbial fuel cells. Bioresource Technology. 2011; 102: 316-323. DOI: 10.1016/j.biortech.2010.06.096

[11] Zaybak Z, Pisciotta JM, Tokash JC, Logan BE. Enhanced start-up of anaerobic facultatively autotrophic biocathodes in bioelectrochemical systems. Journal of Biotechnology. 2013; 168 (4): 478-485. DOI: 10.1016/j.jbiotec.2013.10.001

[12] Estrada, E. B., Salazar, M. d. (2013). Generación de energía eléctrica a partir del tratamiento de aguas residuales por medio de bioceldas. Jiutepec, Morelos, México.: IMTA.

[13] Meng F, Zhao Q, Zheng Z, Wei L, Wang K, Jiang J, Ding J, Na X. (2019). Simultaneous sludge degradation, desalination and bioelectricity generation in two-phase microbial desalination cells. Chemical Engineering Journal. 2018; 361: 180-188.

[14] Metcalf & Eddy, Inc. (2014). Wastewater engineering. Treatment and reuse recovery. 5th ed. United States of America: McGraw Hill; 2014. Print ISBN: 9780073401188. p. 1952.

[15] Yu H, Zhao Q, Dong Q, Jiang J, Wang K, Zhang Y. Electronic and metagenomic insights into the performance of bioelectrochemical reactor simultaneously treating sewage sludge and Cr (VI)-laden wastewater. Chemical Engineering Journal. 2018; 341: 495-504. DOI: 10.1016/j.cej.2018.01.159

[16] Zhang G, Zhao Q, Jiao Y, Wang K, Lee, DJ, Ren N. Efficient electricity generation from sewage sludge using biocathode microbial fuel cell. Water Research. 2012; 46: 43-52. DOI: 10.1016/j.watres.2011.10.036

[17] Huarachi-Olivera R, Dueñas-Gonza A, Yapo-Pari U, Vega P, Romero-Ugarte, M, Tapia J, Molina L, Lazarte-Rivera A, Pacheco-Salazar DG, Esparza M. Bioelectrogenesis with microbial fuel cells (MFCs) using the microalga Chlorella vulgaris and bacterial communities. Electronic Journal of Biotechnology. 2018; 31: 34-43. DOI: 10.1016/j.ejbt.2017.10.013

[18] Feng Y, Yang Q, Wang X, Logan BE. Treatment of carbon fiber brush anodes for improving power generation in air-cathode microbial fuel cells. Journal of Power Sources. 2010; 195: 184-1844. DOI:10.1016/j.jpowsour.2009.10.030

[19] Rabaey K, Ossieur W, Verhaege M, Verstraete W. (2005). Continuous microbial fuel cells convert carbohydrates to electricity. Water Science & Technology. 2005; 52(1-2): 515-523. DOI: 10.2166/wst.2005.0561

[20] Zhao H, Zhao J, Li F, Li X. Performance of denitrifying microbial fuel cell with biocathode over nitrite. Frontiers in Microbiology. 2016; 7(344): 1-7. DOI: 10.3389/fmicb.2016.00344

[21] APHA. Standards methods for the examination of water and wastewater. 23th ed. Washington, DC, USA: American Public Health Asociation; 2017. ISBN: 9780875532875; p. 2076.

[22] Revelo DM, Hurtado NH, Ruiz JO, López S. (2015). Uso de microorganismos nativos en la remoción simultánea de materia orgánica y Cr(VI) en una celda de combustible microbiana de biocátodo (CCM). Información Tecnológica. 2015; 26: 77-88. DOI:

[23] Varanasi JL, Veerubhotla R, Das D. Diagnostic tools for the assessment of MFC. In Debabrata D, Microbial fuel cell: A bioelectrochemical system that converts waste to watts. Kharagpur, India: Springer, 2018. p. 249-262. Ch13.

[24] Al-Mamun A, Jafary T, Baawain MS, Rahman S, Rahman M, Tabatabaei M, Lam SS. Energy recovery and carbon/nitrogen removal from sewage and contaminated groundwater in a coupled hydrolytic-acidogenic sequencing batch reactor and denitrifying biocathode microbial fuel cell. Environmental Research. 2020; 183: 1-11. DOI: 10.1016/j.envres.2020.109273

[25] Freguia S, Rabaey K, Yuan Z, Keller J. Electron and carbon balances in microbial fuel cells reveal temporary bacterial storage behavior during electricity generation. Environmental Science & Technology. 2007; 41: 2915-2921. DOI: 10.1021/es062611i

[26] Min B, Cheng S, Logan BE. Electricity generation using membrane and salt bridge microbial fuel cells. Water Research. 2005; 39(9): 1675-1686. DOI: 10.1016/j.watres.2005.02.002

[27] Tang X, Guo K, Li H, Du Z, Tian J. (2010). Microfiltration membrane performance in two-chamber microbial fuel cells. Biochem. Eng. 2010; 52: 194-198. DOI: 10.1016/j.bej.2010.08.007

[28] Chen S, Patil SA, Schröder U. A high-performance rotating graphite fiber brush air-cathode for microbial fuel cells. Applied Energy Journal. 2018; 211:1089-1094. DOI: 10.1016/j.apenergy.2017.12.013)10.1016/j.apenergy.2017.12.013)

Section 3

Emerging Technologies

Chapter 5

An Overview of Occurrence and Removal of Pharmaceuticals from Sewage/Wastewater

Mohd Salim Mahtab and Izharul Haq Farooqi

Abstract

Nowadays, the occurrence of pharmaceuticals in sewage/wastewater is a major environmental concern. Their precise characterization and suitable treatment/ disposal is a must else it pollutes the surface water bodies and causes major distress on aquatic lives and human health. Also, the up-gradation of the sewage/wastewater treatment plant (WWTP) is a must to consider the removal of these pollutants and to provide the best quality effluent for various reuse purposes. Mostly, the conventional treatment methods are inefficient for their removal, and hence, the most advanced and refined treatment options are needed for their effective treatment. In this chapter, we have highlighted the occurrence of pharmaceuticals in various water samples and their treatment options are reviewed. It was recommended that integrated treatment systems are more efficient, economical, and environmental friendly than single stand-alone treatment. Further advancement and modifications in the treatment options are required to overcome the shortcomings regarding pharmaceutical removal to achieve the legal standard discharge limit.

Keywords: advanced oxidation process, biological treatment, emerging contaminants, wastewater, recalcitrant compounds, sewage

1. Introduction

Nowadays, the problems associated with the widespread occurrence of pharmaceuticals in the aquatic environment have been recognized as an emerging environmental issue [1–3]. The increasing usage of pharmaceuticals and their improper discharge is one of the major environmental concerns. Pharmaceuticals are a large and diverse group of compounds designed to prevent, cure, and treat disease and improve health. Their usage and consumption are increasing consistently due to the discoveries of new drugs, the expanding population, etc. [2, 3]. After intake, these pharmaceutically active compounds undergo metabolic processes in the organism. Significant fractions of the parent compound are excreted in un-metabolized form into raw sewage and wastewater treatment systems. The most commonly occurring pharmaceuticals in the environment are given in **Table 1** [4]. Thus, body metabolization and excretion followed by wastewater treatment are considered to be the primary pathway of pharmaceuticals to the environment [1–3, 5–7]. Disposal of drug leftovers into sewage and trash is another source of entry [8]. In addition, sewer leaking [9], sewer overflow [10], and surface runoff [11] are also considered

S. no.	Class of drugs	Name of drugs
1.	Antibiotics	Erythromycin, ofloxacin, streptomycin, flumequine, ciprofloxacin, trimethoprim, sulfamethoxazole, lincomycin, penicillin, and amoxicillin
2.	Antidepressants	Mianserin
3.	Anticancer drugs	Cyclophosphamide and ifosphamide
4.	Anti-inflammatory drugs	Acetylsalicylic acid (aspirin), diclofenac, ibuprofen, acetaminophen, naproxen, and phenazone
5.	Beta-blockers	Metoprolol, propranolol, nadolol, and atenolol
6.	Diuretics	Furosemide
7.	Lipid regulators	Bezafibrate, gemfibrozil, clofibric acid, and fenofibrate
8.	Steroids and related hormones	17-β-estradiol, estrone, and diethylstilbestrol
9.	Tranquilizers	Diazepam

Table 1.
Some common pharmaceuticals are found in the environment [4].

as additional sources contributing to the presence of pharmaceuticals in the aquatic environment [5].

Their detection techniques and proper characterization are relatively difficult which required distinctive procedures and sophisticated instruments due to their low concentration levels in different environmental matrices [7, 11, 12]. Several studies investigated the occurrence and distribution of pharmaceuticals in soil irrigated with reclaimed water [13, 14] and soil that received biosolids from urban sewage treatment plants [15, 16]. These studies confirmed that the conventional systems are not enough to completely remove such micro-pollutants from wastewater and sludge, and as a result, they find their way into the environment [17]. Once entered the environment, pharmaceutically active compounds can produce subtle effects on aquatic and terrestrial organisms. Therefore, the occurrence of pharmaceutical compounds and the extent to which they can be eliminated during wastewater treatment have become the active subject matter of actual research [1, 3–7].

Domestic sewage is relatively simple to treat with conventional methods due to the absence of any recalcitrant compounds. The conventional treatment options are widely applicable for their effective treatment [1, 18–20]. The sewage/wastewater treatment plants are generally not designed to consider the specific pharmaceuticals, emerging compounds, etc., during the treatment. Hence, their presence in the sewage water is very problematic for the treatment performance of the plant [1, 5–7, 21]. Furthermore, the presence of pharmaceuticals in the effluents of sewage/wastewater treatment plants is very toxic in many ways to the soil and surrounding water bodies [1–5, 21]. To overcome the abovementioned problems, firstly, we have to stop the improper disposal of pharmaceuticals and their proper monitoring/collection system should be designed [3]. The accurate characterization and suitable treatment options should be provided to obtain the legal effluent discharge standards. The constant discharge of various pharmaceuticals into the water bodies and their persistent nature and bioaccumulation potential cause serious effects to aquatic lives and human health [21–23]. Therefore, in this chapter, we have highlighted the occurrence and some of the removal techniques specifically for the pharmaceuticals from sewage/wastewater. The scope for future research directions is also highlighted in the conclusion part.

2. Occurrence of pharmaceuticals in sewage/wastewater

The huge variation in the concentrations of pharmaceutically active compounds (PhACs) was observed due to various factors viz. environmental persistency, dilution, treatment efficiency [21, 24, 25]. In some studies, the reported amounts of pharmaceuticals are estimated to be 5.6, 2.0, and 0.4 g/day/1000 equivalent inhabitants [1, 21]. In one of the studies, the highest levels at the influent of WWTPs were observed for nonsteroidal anti-inflammatory drugs (NSAIDs) that were expected due to their high consumption [1]. Lower but still significant levels of lipid-modifying agents (7–12%), diuretics (8–10%), and beta-blockers (5–9%) were detected entering the WWTPs [1]. Atenolol and carbamazepine were quantified in the influent samples of WWTPs in average concentrations ranging from 0.4 to 1.4 mg/L [1]. The amount found in effluent or sludge depended on the removal efficiency of the plant and/or the physicochemical properties of the compounds. In the effluent waters, NSAIDs were present in the highest percentage (35–44%), followed by the lipid-modifying agents (8–29%) and psychiatric drugs (17–30%) [1]. The highest concentrations in the effluents were found for naproxen, diclofenac, and carbamazepine [1].

It has been reported that from the list of detected samples of the emerging contaminants about 70% are PACs and personal care products (PCPs) [26]. Globally, more than 200 PhACs have been reported in river waters with a maximum concentration of 6 mg/L for ciprofloxacin antibiotics [27]. Similarly, tamoxifen was detected in the range of 25–38 ng/L [28]. Also, the concentrations of antibiotics, hormones, antidepressants, and chemotherapy drugs range from 0.04 to 6.3 μg/L [29]. Out of the various sources of the PhACs into the environment, the domestic discharge and effluents of the manufacturing units of pharmaceuticals are well-thought-out major sources [22]. Various categories for the occurrence of the PhACs have been reported viz. wastewater treatment plants (WWTPs), wastewater,

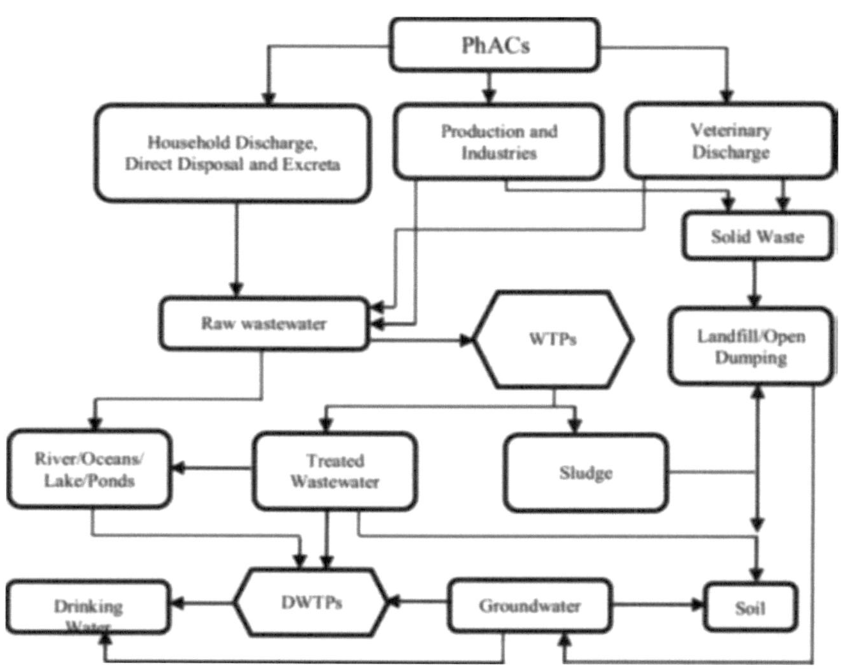

Figure 1.
Flowchart showing PhACs pathways in the environment [30, 31].

sewage, sewage sludge, groundwater, surface water, and drinking water [24, 30, 31]. So, WWTPs are considered as one of the prominent anthropogenic sources emitting pharmaceuticals into the environment along with industrial discharges, hospital effluents, etc. [22, 25]. Furthermore, the inefficient management and treatments of PhACs risk the prospect of sustainable reuse of treated wastewater and sludge [21]. **Figure 1** shows the flowchart showing PhACs pathways in the environment [30, 31].

3. Removal of pharmaceuticals from sewage/wastewater

It was well recognized in the literature that the conventional biological treatment systems alone are not sufficient enough to completely remove the pharmaceuticals and therefore, some additional steps are required for their proper treatment. It was reported that among the conventional activated sludge process (ASP) and membrane bioreactor (MBR) systems, the MBR system appeared to have a higher removal efficiency for many of the pharmaceuticals [1, 32]. On the other hand, carbamazepine and hydrochlorothiazide showed poor removal efficiencies in either ASP or MBR systems [1, 32]. O'Brien et al. [33] have found that compounds such as atenolol, carbamazepine, and ibuprofen appeared to be persistent in the sewer system. It was recommended to use a composite sampling approach in wastewater treatment plants [5, 33]. The superior performance of MBR in the removal of some target pharmaceuticals is due to the result of the higher biomass concentration, longer solid retention time (SRT), and better-retaining capacity of solids and microbes [32, 34]. On the other hand, the integration approach of membrane technology is called electrochemical membrane bioreactors (EMBR). It was observed that EMBRs are more efficient with low energy consumptions as compared to MBRs and ASPs [21, 35]. But the common problem with the advanced technologies is their limited applications only at laboratory and pilot scales. Besides, membrane fouling, high energy demand, and costly membrane materials are some limitations of MBRs, which need to be overcome for their extensive full-scale applications [21].

The term removal of pharmaceuticals used here means the conversion of the parent compound. Thus, the overall removal refers to the losses of a parent compound by different mechanisms of chemical and physical transformation, biodegradation, and sorption to solid matter [1]. The most analyzed carbamazepine showed very low removal (<25%) regardless of the treatment applied [32]. The pharmaceuticals removal efficiencies are based on the characteristics of the wastewater, treatment types used, and other operational conditions [1, 21]. The addition of the occasional tertiary treatment improves the removal efficiencies of the pharmaceuticals. The lower removal efficiency of diclofenac was reported in some studies [1, 36, 37]. Better performances of WWTP may be due to longer both hydraulic and solid retention times. As a compound spends more time in reactors wherein bacteria growth is promoted, the biological transformation may occur to a greater extent [38]. It has been proven that longer SRT, especially, improves the elimination of most of the pharmaceuticals during sewage treatment [1, 39].

A variety of treatment techniques for pharmaceuticals removal have been considered in the past studies such as natural, conventional and advanced treatment approaches. Dilution, volatilization, photolysis, sorption, biodegradation, etc., are cost-effective and natural processes [21]. However, the natural processes are proved less efficient [22]. On the other hand, the conventional approaches viz. adsorption, ozonation, membrane filtration, showed high pharmaceuticals removal efficacies [23]. But these approaches are having some disadvantages like oxidation by-products formation in the ozonation process may be more toxic than the parent compounds, and high operational costs in addition to the concentrate disposal are required in the membrane filtration process [25]. The widespread applications of various advanced treatment approaches viz. advanced oxidation processes (AOPs), constructed wetlands, bioelectrical systems, enzymatic treatment, have been recommended in the past few years [21]. Also, the up-gradation of the conventional WWTPs might further minimize the environmental release of the various pharmaceuticals [21, 23, 40]. Although the AOPs are considered one of the most effective treatment options for a variety of pharmaceuticals removal, their full-scale applications are still limited due to the number of challenges [18–21, 25, 41].

The WWTPs generally considered the primary, secondary, and sometimes tertiary treatment stages. The pharmaceuticals entered into the plants undergo several treatment stages, and their fraction is degraded/removed [21, 24, 42]. In the secondary stage, the pharmaceuticals are subjected to several processes such as biodegradation, sorption, dispersion, dilution, photodegradation, and volatilization [21, 22, 24]. Likewise, the tertiary treatment steps are reported to exhibit significant pharmaceuticals removal efficiency via ozonation-like conventional oxidation processes [21, 43, 44].

The importance of the tertiary treatment in the WWTPs is versatile as it supplements the secondary treatment and those pollutants that are not removed in the second stage are removed in the tertiary stage. Several advanced technologies are employed to remove the pharmaceuticals in the WWTPs themselves to produce high-quality effluent for reuse purposes [21, 44, 45]. Among the tertiary treatment, AOPs have been considered that oxidize/mineralize the various pharmaceuticals and their by-products to CO_2, H_2O, and simple inorganic ions [18, 21]. The various types are AOPs are now widely applied for various applications of high strength and pharmaceuticals removal viz. Fenton process, Photo-Fenton process, Electro-Fenton process, Sono-Fenton process, ozonation process, UV-based treatment [21, 46]. Also, a range of commercially available adsorbents, such as activated carbon (AC), biochar, carbon nanotubes, clay minerals, are used for the adsorption of various pharmaceuticals [21, 47]. The usages of AC for a broad-spectrum pharmaceutical adsorption were found most suitable due to reduced interference from the organic materials for the adsorption active sites [21, 48]. The adsorption efficiency depends on the types of PhACs, properties of AC, and other environmental conditions [21, 24].

Among the mentioned options, ozonation and AC treatment are found to be the economically feasible option and utilized in some WWTPs [21, 25]. The main reactive species in AOPs for the degradation/mineralization of the pharmaceuticals are hydroxyl radicals ($OH^{\bullet}$) and the number of parallel reactions is reported in their mechanism [18–20, 49, 50]. The suitability of the various adoption of the AOPs is mainly based on wastewater characteristics, recalcitrant nature of the target compounds, available resources, and economic conditions [50]. It was well recognized in the literature that the integrated processes are more efficient and environmental friendly [18, 50]. A very high removal efficiency (>95%) of diclofenac, carbamazepine, sulpiride, at an ozone dose of 5 mg/L, was observed [51]. All the AOPs are having their limitations/disadvantages as well; hence, the suitable/optimized treatment options should be designed and implemented to achieve the target removal efficacies etc. [18, 41, 50]. Some of the disadvantages of the Fenton process are low-working pH requirement and high sludge production, the chances of the pharmaceuticals accumulate in the iron sludge produced after the treatment [41]. On the other hand, when the applied ozone dosages are inadequate, it will result in the formation of transformation products [18], and the toxicity can further be reduced by a subsequent biological treatment [21, 52]. The combined approach of the ozonation-biological process is found most efficient for the removal of pharmaceuticals from secondary urban wastewater [21, 52, 53]. Currently, many treatment technologies are available as mentioned in **Table 2** [54, 55].

Treatment technologies	Classification
Physical treatment	Primary treatment
Aerobic process Anaerobic process	Secondary biological treatment
Activated carbon Membrane distillation Membrane technology	Tertiary treatments
Fenton process Ozone/hydrogen peroxide treatment Photocatalysis Electrochemical oxidation Ultrasound irradiation Wet air oxidation	Advanced oxidation processes
Mixed primary, secondary and tertiary treatments	Hybrid technologies

Table 2.
Some treatment technologies for pharmaceutical wastewater treatment [54, 55].

4. Conclusion

This chapter provides a brief overview regarding the problems associated with the pharmaceuticals present in the sewage/wastewater and their suitable treatment options. From the literature, it was understood that the problems related to the emerging contaminants and particularly for the pharmaceuticals are of great concern and require specific attention to protecting the environment and public health. Out of the various categories of pharmaceuticals, different treatment options are required and one single option is not sufficient to remove all the types of pharmaceuticals. The challenges associated with their accurate analysis, detection, and extraction due to their low concentration are also an important domain for further research. Regarding the treatment options in various studies, it was reported that the integrated processes are more advantageous in many ways for pharmaceuticals removal. For example, the post-biological treatment option after the ozonation process significantly improves the pharmaceutical removal. The other options like ASP and MBR are also considered useful but not efficient enough for their complete mineralization and removal. Also, the activated carbon-adsorption process is just a phase change mechanism system and required extensive research for further improvement. Various transformation/intermediate products are formed in AOPs treatment, hence required more advancements to remove those toxic intermediates from the water matrix. The up-gradation of the WWTPs is a very important step to improve the effluent quality considering the problems of the pharmaceuticals. The single and combined AOPs are limited to lab/pilot scale only and their full-scale applications are required, which should be focused on in future research for the best-fit alternative both economically and environmental friendly.

Acknowledgements

The authors are grateful to the editor and reviewers.

Conflict of interest

No potential conflict of interest was reported by the authors.

Author details

Mohd Salim Mahtab* and Izharul Haq Farooqi
Department of Civil Engineering, Z.H. College of Engineering and Technology, Aligarh Muslim University, Aligarh, India

*Address all correspondence to: msmahtab@zhcet.ac.in

References

[1] Jelic A, Gros M, Ginebreda A, Cespedes-Sànchez R, Ventura F, Petrovic M, et al. Occurrence, partition and removal of pharmaceuticals in sewage water and sludge during wastewater treatment. Water Research. 2011;**45**(3):1165-1176. DOI: 10.1016/j.watres.2010.11.010

[2] Husain Khan A, Abdul Aziz H, Khan NA, Ahmed S, Mehtab MS, Vambol S, et al. Pharmaceuticals of emerging concern in hospital wastewater: Removal of Ibuprofen and Ofloxacin drugs using MBBR method. International Journal of Environmental Analytical Chemistry. 2020;**19**:1-5. DOI: 10.1080/03067319.2020.1855333

[3] Khan AH, Aziz HA, Khan NA, Hasan MA, Ahmed S, Farooqi IH, et al. Impact, disease outbreak and the eco-hazards associated with pharmaceutical residues: A Critical review. International Journal of Environmental Science and Technology. 2021;**10**:1-2. DOI: 10.1007/s13762-021-03158-9

[4] Nikolaou A, Meric S, Fatta D. Occurrence patterns of pharmaceuticals in water and wastewater environments. Analytical and Bioanalytical Chemistry. 2007;**387**(4):1225-1234. DOI: 10.1007/s00216-006-1035-8

[5] Tran NH, Gin KY. Occurrence and removal of pharmaceuticals, hormones, personal care products, and endocrine disrupters in a full-scale water reclamation plant. Science of the Total Environment. 2017;**599**:1503-1516. DOI: 10.1016/j.scitotenv.2017.05.097

[6] Madikizela LM, Ncube S, Chimuka L. Analysis, occurrence and removal of pharmaceuticals in African water resources: A current status. Journal of Environmental Management. 2020;**253**:109741. DOI: 10.1016/j.jenvman.2019.109741

[7] Mousel D, Bastian D, Firk J, Palmowski L, Pinnekamp J. Removal of pharmaceuticals from wastewater of health care facilities. Science of the Total Environment. 2021;**751**:141310. DOI: 10.1016/j.scitotenv.2020.141310

[8] Ruhoy IS, Daughton CG. Beyond the medicine cabinet: An analysis of where and why medications accumulate. Environment International. 2008;**34**(8):1157-1169. DOI: 10.1016/j.envint.2008.05.002

[9] Wolf L, Zwiener C, Zemann M. Tracking artificial sweeteners and pharmaceuticals introduced into urban groundwater by leaking sewer networks. Science of the Total Environment. 2012;**430**:8-19. DOI: 10.1016/j.scitotenv.2012.04.059

[10] Launay MA, Dittmer U, Steinmetz H. Organic micropollutants discharged by combined sewer overflows–characterisation of pollutant sources and stormwater-related processes. Water Research. 2016;**104**:82-92. DOI: 10.1016/j.watres.2016.07.068

[11] Pedersen JA, Soliman M, Suffet IH. Human pharmaceuticals, hormones, and personal care product ingredients in runoff from agricultural fields irrigated with treated wastewater. Journal of Agricultural and Food Chemistry. 2005;**53**(5):1625-1632. DOI: 10.1021/jf049228m

[12] Majumder A, Gupta AK, Ghosal PS, Varma M. A review on hospital wastewater treatment: A special emphasis on occurrence and removal of pharmaceutically active compounds, resistant microorganisms, and SARS-CoV-2. Journal of Environmental Chemical Engineering. 2020;**22**:104812. DOI: 10.1016/j.jece.2020.104812

[13] Gielen GJ, van den Heuvel MR, Clinton PW, Greenfield LG. Factors

impacting on pharmaceutical leaching following sewage application to land. Chemosphere. 2009;**74**(4):537-542. DOI: 10.1016/j.chemosphere.2008.09.048

[14] Kinney CA, Furlong ET, Werner SL, Cahill JD. Presence and distribution of wastewater-derived pharmaceuticals in soil irrigated with reclaimed water. Environmental Toxicology and Chemistry: An International Journal. 2006;**25**(2):317-326. DOI: 10.1897/05-187R.1

[15] Carbonell G, Pro J, Gómez N, Babín MM, Fernández C, Alonso E, et al. Sewage sludge applied to agricultural soil: Ecotoxicological effects on representative soil organisms. Ecotoxicology and Environmental Safety. 2009;**72**(4):1309-1319. DOI: 10.1016/j.ecoenv.2009.01.007

[16] Lapen DR, Topp E, Metcalfe CD, Li H, Edwards M, Gottschall N, et al. Pharmaceutical and personal care products in tile drainage following land application of municipal biosolids. Science of the Total Environment. 2008;**399**(1-3):50-65. DOI: 10.1016/j.scitotenv.2008.02.025

[17] Wick A, Fink G, Joss A, Siegrist H, Ternes TA. Fate of beta blockers and psycho-active drugs in conventional wastewater treatment. Water Research. 2009;**43**(4):1060-1074. DOI: 10.1016/j.watres.2008.11.031

[18] Hussain M, Mahtab MS, Farooqi IH. The applications of ozone-based advanced oxidation processes for wastewater treatment: A review. Advances in Environmental Research. 2020;**9**(3):191-214. DOI: 10.12989/aer.2020.9.3.191

[19] Mahtab MS, Farooqi IH, Khursheed A. Sustainable approaches to the Fenton process for wastewater treatment: A review. Materials Today: Proceedings. 2021;**47**(7):1480-1484. DOI: 10.1016/j.matpr.2021.04.215

[20] Hussain M, Mahtab MS, Farooqi IH. A comprehensive review of the Fenton-based approaches focusing on landfill leachate treatment. Advances in Environmental Research. 2021;**10**(1):59-86. DOI: 10.12989/aer.2021.10.1.059

[21] Rout PR, Zhang TC, Bhunia P, Surampalli RY. Treatment technologies for emerging contaminants in wastewater treatment plants: A review. Science of the Total Environment. 2021;**753**:141990. DOI: 10.1016/j.scitotenv.2020.141990

[22] Barbosa MO, Moreira NF, Ribeiro AR, Pereira MF, Silva AM. Occurrence and removal of organic micropollutants: An overview of the watch list of EU Decision 2015/495. Water Research. 2016;**94**:257-279. DOI: 10.1016/j.watres.2016.02.047

[23] Pesqueira JF, Pereira MF, Silva AM. Environmental impact assessment of advanced urban wastewater treatment technologies for the removal of priority substances and contaminants of emerging concern: A review. Journal of Cleaner Production. 2020;**261**:121078. DOI: 10.1016/j.jclepro.2020.121078

[24] Luo Y, Guo W, Ngo HH, Nghiem LD, Hai FI, Zhang J, et al. A review on the occurrence of micropollutants in the aquatic environment and their fate and removal during wastewater treatment. Science of the Total Environment. 2014;**473**:619-641. DOI: 10.1016/j.scitotenv.2013.12.065

[25] Rizzo L, Malato S, Antakyali D, Beretsou VG, Đolić MB, Gernjak W, et al. Consolidated vs new advanced treatment methods for the removal of contaminants of emerging concern from urban wastewater. Science of the Total Environment. 2019;**655**:986-1008. DOI: 10.1016/j.scitotenv.2018.11.265

[26] Das S, Ray NM, Wan J, Khan A, Chakraborty T, Ray MB. Micropollutants in wastewater: Fate and

removal processes. Physico-chemical Wastewater Treatment and Resource Recovery. 2017;**3**:75

[27] Petrie B, Barden R, Kasprzyk-Hordern B. A review on emerging contaminants in wastewaters and the environment: Current knowledge, understudied areas and recommendations for future monitoring. Water Research. 2015;**72**:3-27. DOI: 10.1016/j.watres.2014.08.053

[28] Ferrando-Climent L, Rodriguez-Mozaz S, Barceló D. Incidence of anticancer drugs in an aquatic urban system: From hospital effluents through urban wastewater to natural environment. Environmental Pollution. 2014;**193**:216-223. DOI: 10.1016/j.envpol.2014.07.002

[29] Jones OA, Voulvoulis N, Lester JN. Human pharmaceuticals in the aquatic environment: A review. Environmental Technology. 2001;**22**(12):1383-1394. DOI: 10.1080/09593332208618186

[30] Balakrishna K, Rath A, Praveenkumarreddy Y, Guruge KS, Subedi B. A review of the occurrence of pharmaceuticals and personal care products in Indian water bodies. Ecotoxicology and Environmental Safety. 2017;**137**:113-120. DOI: 10.1016/j.ecoenv.2016.11.014

[31] Khan SU, Rameez H, Basheer F, Farooqi IH. Eco-toxicity and health issues associated with the pharmaceuticals in aqueous environments: A global scenario. Pharmaceutical Wastewater Treatment Technologies: Concepts and Implementation Strategies. 2021:145-179. DOI: 10.2166/9781789061338

[32] Radjenović J, Petrović M, Barceló D. Fate and distribution of pharmaceuticals in wastewater and sewage sludge of the conventional activated sludge (CAS) and advanced membrane bioreactor (MBR) treatment. Water Research. 2009;**43**(3):831-841. DOI: 10.1016/j.watres.2008.11.043

[33] O'Brien JW, Banks AP, Novic AJ, Mueller JF, Jiang G, Ort C, et al. Impact of in-sewer degradation of pharmaceutical and personal care products (PPCPs) population markers on a population model. Environmental Science & Technology. 2017;**51**(7):3816-3823. DOI: 10.1021/acs.est.6b02755

[34] Joss A, Zabczynski S, Göbel A, Hoffmann B, Löffler D, McArdell CS, et al. Biological degradation of pharmaceuticals in municipal wastewater treatment: Proposing a classification scheme. Water Research. 2006;**40**(8):1686-1696. DOI: 10.1016/j.watres.2006.02.014

[35] Asif MB, Maqbool T, Zhang Z. Electrochemical membrane bioreactors: State-of-the-art and future prospects. Science of the Total Environment. 2020;**741**:140233

[36] Cirja M, Ivashechkin P, Schäffer A, Corvini PF. Factors affecting the removal of organic micropollutants from wastewater in conventional treatment plants (CTP) and membrane bioreactors (MBR). Reviews in Environmental Science and Bio/Technology. 2008;**7**(1):61-78. DOI: 10.1007/s11157-007-9121-8

[37] Kimura K, Hara H, Watanabe Y. Elimination of selected acidic pharmaceuticals from municipal wastewater by an activated sludge system and membrane bioreactors. Environmental Science & Technology. 2007;**41**(10):3708-3714. DOI: 10.1021/es061684z

[38] Reif R, Suárez S, Omil F, Lema JM. Fate of pharmaceuticals and cosmetic ingredients during the operation of a MBR treating sewage. Desalination. 2008;**221**(1-3):511-517. DOI: 10.1016/j.desal.2007.01.111

[39] Göbel A, McArdell CS, Joss A, Siegrist H, Giger W. Fate of sulfonamides, macrolides, and trimethoprim in different wastewater treatment technologies. Science of the Total Environment. 2007;**372**(2-3):361-371. DOI: 10.1016/j.scitotenv.2006.07.039

[40] Roccaro P. Treatment processes for municipal wastewater reclamation: The challenges of emerging contaminants and direct potable reuse. Current Opinion in Environmental Science & Health. 2018;**2**:46-54. DOI: 10.1016/j.coesh.2018.02.003

[41] Mahtab MS, Farooqi IH, Khursheed A. Zero Fenton sludge discharge: A review on reuse approach during wastewater treatment by the advanced oxidation process. International Journal of Environmental Science and Technology. 2021;**13**:1-4. DOI: 10.1007/s13762-020-03121-0

[42] Tran NH, Reinhard M, Gin KY. Occurrence and fate of emerging contaminants in municipal wastewater treatment plants from different geographical regions—A review. Water Research. 2018;**133**:182-207. DOI: 10.1016/j.watres.2017.12.029

[43] Rout PR, Bhunia P, Dash RR. Response surface optimization of phosphate removal from aqueous solution using a natural adsorbent. Trends in Asian Water Environmental Science and Technology. 2017:93-104. DOI: 10.1007/978-3-319-39259-2_8

[44] Ahmed MB, Zhou JL, Ngo HH, Guo W, Thomaidis NS, Xu J. Progress in the biological and chemical treatment technologies for emerging contaminant removal from wastewater: A critical review. Journal of Hazardous Materials. 2017;**323**:274-298. DOI: 10.1016/j.jhazmat.2016.04.045

[45] Wang J, Wang S. Removal of pharmaceuticals and personal care products (PPCPs) from wastewater: A review. Journal of Environmental Management. 2016;**182**:620-640. DOI: 10.1016/j.jenvman.2016.07.049

[46] de Oliveira M, Frihling BE, Velasques J, Magalhães Filho FJ, Cavalheri PS, Migliolo L. Pharmaceuticals residues and xenobiotics contaminants: Occurrence, analytical techniques and sustainable alternatives for wastewater treatment. Science of the Total Environment. 2020;**705**:135568. DOI: 10.1016/j.scitotenv.2019.135568

[47] Rodriguez-Narvaez OM, Peralta-Hernandez JM, Goonetilleke A, Bandala ER. Treatment technologies for emerging contaminants in water: A review. Chemical Engineering Journal. 2017;**323**:361-380. DOI: 10.1016/j.cej.2017.04.106

[48] Budimirović D, Veličković ZS, Djokić VR, Milosavljević M, Markovski J, Lević S, et al. Efficient As (V) removal by α-FeOOH and α-FeOOH/α-MnO_2 embedded PEG-6-arm functionalized multiwall carbon nanotubes. Chemical Engineering Research and Design. 2017;**119**:75-86. DOI: 10.1016/j.cherd.2017.01.010

[49] Mahtab MS, Farooqi IH. UV-TiO_2 process for landfill leachate treatment: Optimization by response surface methodology. International Journal for Research in Engineering Application & Management. 2020;**5**(12):14-18. DOI: 10.35291/2454-9150.2020.0160

[50] Mahtab MS, Islam DT, Farooqi IH. Optimization of the process variables for landfill leachate treatment using Fenton based advanced oxidation technique. Engineering Science and Technology, an International Journal. 2021;**24**(2):428-435. DOI: 10.1016/j.jestch.2020.08.013

[51] Sui Q, Huang J, Deng S, Yu G, Fan Q. Occurrence and removal of

pharmaceuticals, caffeine and DEET in wastewater treatment plants of Beijing, China. Water Research. 2010;**44**(2):417-426. DOI: 10.1016/j.watres.2009.07.010

[52] Knopp G, Prasse C, Ternes TA, Cornel P. Elimination of micropollutants and transformation products from a wastewater treatment plant effluent through pilot scale ozonation followed by various activated carbon and biological filters. Water Research. 2016;**100**:580-592. DOI: 10.1016/j.watres.2016.04.069

[53] McArdell C. The first full-scale advanced ozonation plant in the Dübendorf WWTP running; the new Swiss water protection act approved. Norman Bulletin. 2015;**4**:36-37

[54] Mahmood Q, Khan MS, Riaz N. Existing treatment—Globally in full scale plants. Pharmaceutical Wastewater Treatment Technologies: Concepts and Implementation Strategies. 2021:402-428. DOI: 10.2166/9781789061338

[55] Khan NA, Ahmed S, Vambol V, Vambol S. Pharmaceutical Wastewater Treatment Technologies: Concepts and Implementation Strategies2021. DOI: 10.2166/9781789061338

Chapter 6

Implications of Sewage Discharge on Freshwater Ecosystems

Sami Ullah Bhat and Umara Qayoom

Abstract

Freshwater ecosystems such as lakes and rivers are among the sensitive ecosystems, which host rich biodiversity. Being major freshwater resources, they provide a wide range of ecosystem services, making their existence essential for the well-being of human societies. However, in the past few decades, there have been adverse impacts on the health of these ecosystems due to uncontrolled sewage disposal throughout the world. This is increasingly becoming a tough challenge to protect the freshwater ecosystems from the ramifications of the entry of untreated sewage. Loss of biodiversity, physiological and behavioral changes in species, community shifts, and fish mortality have been witnessed in aquatic ecosystems, which are the recipients of untreated or partially treated sewage. Nutrients such as nitrogen and phosphorus are abundant in sewage and are one of the leading causes of eutrophication of water bodies. Several freshwater ecosystems around the world have become a victim of eutrophication due to untreated sewage disposal, leading to a change in trophic status.

Keywords: biodiversity, eutrophication, phosphorus, species

1. Introduction

One of the biggest challenges we are confronting in the twenty-first century is the inaccessibility of clean water and improved sanitation [1]. Although safe drinking water, sanitation and hygiene (WASH) are required for an improved standard of living, they are equally important for the protection of health and environment. As the countries improvise their sanitation coverage, it is also important that they reduce the release of untreated sewage into the environment by exploiting the energy and nutrients present in it. Water plays an important role in various aspects of socio-economic development such as food production, economy, domestic water supply, environmental sustainability, health systems, and industrial applications. Lack of access to WASH could have a negative impact on the economy, health, and environment. Water pollution from sources such as agriculture, industries, urban runoffs, and waste disposal threatens clean drinking water supplies with detrimental impacts on freshwater ecosystems [2]. Several water bodies in developing countries such as rivers, streams, and lakes, which are located close to highly populated areas, have become filled with waste, which have turned them into dead or sewage streams. Most aquatic ecosystems have a natural tendency to dilute pollution to some extent, but severe contamination of aquatic ecosystems results in the alteration of their fauna and flora community [3]. The amount of nutrients received by aquatic ecosystems varies throughout the world depending on the characteristics of the ecosystem. Most of them receive varying quantities of a wide range of nutrients that are unloaded from human settlements. Sewage is the used water-containing solids deposited from households, commercials, and industries, which is transported in sewers and disposed off into watercourses. Sewage disposal in a particular region depends on the accessibility of natural watercourses in that particular area. Worldwide around 65% of the river stretches are polluted [4], which have resulted in the degradation as well as loss of biodiversity in the water bodies and cannot be neglected.

Poorly managed human excreta has several detrimental consequences on the environment, polluting surface water such as lakes and rivers. Heavily polluted water has a serious impact on freshwater ecosystems, food webs, and biodiversity. Water bodies located in highly populated urban areas have a considerable amount of biological oxygen demand contributed mostly from wastewater. Untreated or partly treated urban wastewater consists of high concentration of nutrients as well as organic matter [5], which upon decomposition releases additional nutrients. Increased levels of nutrients especially nitrogen and phosphorus in aquatic ecosystems are associated with eutrophication. Algal blooms especially those of cyanophytes release cyanotoxins [6], which are known to have harmful effects on aquatic life, wildlife, livestock, agricultural crops, and humans [7]. Several toxins are liberated from sewage into the water, which are consumed by fishes and other forms of aquatic life thereby increasing their possibility of entering into the food chain. Several toxic substances including heavy metals have a high concentration in the wastewater generated from industries [8]. Due to their non-degradable nature, they tend to display high toxicity in aquatic systems and accumulate in the food webs. Thus, water pollution has received more attention during the past few years owing to their ecologic, biodiversity, economic, and social perspectives.

2. Composition of sewage

Sewage is composed of domestic effluent consisting of black water (excreta, urine, and fecal sludge) and gray water (used water from washing and bathing); water from commercial establishments and institutions, including hospitals; industrial effluent, stormwater and other urban runoff; and agricultural, horticultural and aquaculture runoff [9]. Sewage comprises 99.9% of water and 0.1% of solids, which includes dissolved as well as suspended organic and inorganic solids. Dissolved solids constitute a major portion as compared to suspended solids, while organic fraction consists of fats, carbohydrates, proteins, lignin, and their decomposition products. Similarly, the inorganic part includes several constituents derived from industrial as well as domestic sources including heavy metals like cadmium, mercury, arsenic, zinc, and copper. A varied and abundant diversity of microbes are present in sewage [10] and is contributed in human sewage from sources like human domestic waste such as feces, washing, bathing urine, and sweat. These microorganisms are added from the human body present in the skin, respiratory tract, oral cavity, gastrointestinal tract, and urogenital tract. Wastewater is considered as an important reservoir of pathogens, [11, 12], which includes Fecal Coliform, *E coli,* Salmonella, Shigella, Vibrio cholera, parasitic eggs and cysts, viruses and fungi [13, 14], intestinal nematodes like hookworm (Ancylostoma duodenale), roundworm (Ascaris lumbricoides), and whipworms (Trichuris). Sewage contains a rich concentration of nutrients such as nitrogen (N) and phosphorus (P). Around, 16.6 Tg (Tg = million metric ton) of nitrogen and 3.0 Tg of phosphorus are present in wastewater produced throughout the world annually. Human urine consists of a high concentration of nutrients such as nitrogen and phosphorus than the amount present in the feces [15]. Thus, human urine is responsible for 50% of the phosphorus load and 80% of the nitrogen load in the sewage [16].

3. Sewage generation and treatment scenario throughout the world

Despite being a major factor in water quality, proper wastewater treatment is lacking in many developing countries of the world and it is estimated that 90% of developing countries do not treat wastewater before disposal into the receiving waters be it lakes or rivers. A large amount of untreated industrial, agricultural, and domestic wastes is discharged into the world's waterways due to which low-income countries are hit by contaminated water supplies and disease. Water sources such as rivers, lakes, and oceans are the major recipients of domestic and industrial wastes across the world. Country-specific data on wastewater generation reveal that around 390 billion m^3 of wastewater is generated throughout the world, which is fivefold the amount of water released by Niagara Falls annually [17]. However, the amount of wastewater generated is projected to increase by 24% by 2020 and further about 51% by 2050 [18]. It is estimated that worldwide about 80% of the wastewater generated is discharged directly without any proper treatment into water bodies [19]. Around two million tons of sewage, industrial and agricultural waste is discharged into the world's waterways. Water sources such as rivers, lakes, and oceans are the major recipients of domestic and industrial wastes across the world. With regard to the wastewater generation and treatment in the world, it has been stated that high-income countries on an average treat around 70% of the wastewater generated by them, and upper- and lower middle-income countries provide treatment to 38% and 28% of the wastewater generated respectively, while as low-income countries treat only 8% of total wastewater generated [20]. Available figures reveal a shocking scenario of sewage disposal into water bodies across the world. Venezuela discharges 97% of sewage generated directly into the environment without providing any treatment. Developed nations like Turkey discharge 75% of the wastewater generated from industries directly into the environment. About 71% of the wastewater generated in European countries receives treatment owing to public awareness toward health and environmental or partly due to technological advancements. In Latin American countries, only 20% of the wastewater generated is being treated, while the rest is disposed off untreated into the water bodies. The Middle East and North Africa provide treatment to 51% [20], while Asian countries treat only 32% of generated wastewater. Asia is the largest producer of wastewater and generates 42% of wastewater produced globally with an annual estimate of 159 billion m^3. North American countries generate 67 billion m^3, while Europe generates 68 billion m^3 [18]. With the ongoing freshwater crisis throughout the world, the available freshwater resources cannot be polluted and made unfit for human consumption. Proper treatment and disposal of wastewater should be regarded as a matter of urgency throughout the world [21].

4. Effects of sewage on freshwater biodiversity

Now a day's there is an increasing recognition that freshwater is a valuable resource due to overexploitation and pollution. Wastewater discharge contain several harmful substances or chemicals, which may cause adverse environmental impacts such as changes in aquatic habitats, species composition, and decrease in biodiversity. All of these impacts lead to a less valuable environment, a less prosperous economy, and ultimately, a diminished quality of life. Several substances are present in sewage, which can potentially impact plant and animal communities in different ways.

4.1 Temperature

4.1.1 Physical changes

Sewage discharge is often associated with physical changes in water bodies. Aquatic life sustains under an optimum temperature and an increase in the average temperature of the water body has ecological impacts resulting in thermal enhancement [22]. The shift in water temperature can seriously affect aquatic life, such as microbes, invertebrates, algae, and fish [23]. Temperature also affects the solubility and consequently, the availability of oxygen in the water. An increase in temperature results in less dissolution of oxygen in the water and hence, oxygen demand required by the bacteria for the degradation of wastes also increases. Tissue anoxia can occur at higher lethal temperatures in aquatic animals. Temperature also affects key physicochemical conditions such as oxygen concentrations as well as energetic processes associated with primary production and litter decomposition [24, 25].

4.1.2 Chemical changes

The effects of certain toxic substances like copper that increases metabolic demand or zinc which blocks oxygen uptake at the gill level for fish get enhanced by an increase in temperature. Toxicants that act on cellular enzymes involved in energy metabolism or that cause a change in the rate of uptake may also have their effect potentiated by a temperature increase. High water temperature also affects the toxicity of some chemicals in the water as well as the sensitivity of living organisms to toxic substances [26, 27].

4.1.3 Biological changes

Causes of thermal death include failure of osmoregulatory processes, alterations in cellular enzymes and membrane lipids, and protein denaturation. In addition, temperature controls the growth rates of phytoplankton, macrophytes, and epiphytes, making freshwater ecosystems sensitive to rising temperatures [28, 29]. Because most river organisms are ectotherms, changes in temperature have profound effects on their growth, phenology, survival, and distribution [30, 31].

4.2 Dissolved oxygen

4.2.1 Physical changes

Dissolved oxygen (DO) is a key parameter that determines the water quality as well as the health of an aquatic ecosystem. The presence of a certain amount of DO in water is important for the survival of higher forms of biodiversity [32]. A fluctuation in DO near its saturation is an indication of relatively healthy waters while as low dissolved oxygen indicates potential danger to the water body [33]. Oxygen-demanding wastes in the sewage are responsible for the depletion of DO levels, which impact both water quality and biodiversity in the water body [34]. The aquatic ecosystem suffering from hypoxic or anoxic conditions is responsible for the depletion of fish stocks and other forms of aquatic life. These losses can have harmful effects on ecological health, economy, and stability of the ecosystem [35].

4.2.2 Chemical changes

Development of hypoxic and anoxic conditions, increase in metal and phosphate release from sediments creation of hypoxic (reduced dissolved oxygen), anoxic (extremely low or no dissolved oxygen) and euxinic (sulfide production in the absence of oxygen) conditions take place in the water body [36, 37]. Low DO level affects the metabolic processes of species. Low levels of DO in receiving water bodies can result in the release of toxic substances, biomagnification in organisms, and increased nutrient loads.

4.2.3 Biological changes

Fish are among the most affected species as low DO concentration increases their susceptibility toward diseases, retarding their growth, hindering their swimming ability, changes in feeding habits, migration, and in extreme cases results in death.

It significantly affect mortality, reproduction, behavior, and physiological response in fishes [38]. If the decrease in oxygen continues for a long time, it can result in a change in species composition [39]. Among planktonic organisms most likely to suffer mortality from exposure to low oxygen in bottom waters are fish larvae lacking fully developed sensory and motor capabilities.

4.3 Total suspended solids

4.3.1 Physical changes

Suspended solids (SS) comprise of a fine particulate matter having a diameter of less than 62 mm. They can cause physical damage to fish gills [40]. Blockage of filter-feeding apparatus of zooplankton, gills of the most sensitive benthic invertebrates like epibenthos, living on or above the sediment can be clogged by sediment particles.

4.3.2 Chemical changes

They can pose a number of direct as well as indirect environmental impacts like reduced sunlight penetration, which in turn affects photosynthesis, toxic effects due to contaminants attached with suspended solids [22]. A high concentration of salts can result in increase in the salt content of the water body with harmful effects on aquatic organisms and a brackish, salty taste to its consumers.

4.3.3 Biological changes

A high level of SS in receiving water body can cause flocculation and sinking of phytoplankton, reduced primary productivity in macrophytes and algae [41, 42], egg mortality in fish [43]. Further, SS in zooplankton can cause toxicity, as well as ingestion of sediment particles having no nutritional value, causing zooplankton starvation and death.

4.4 Cyanide

This substance is an important toxicant to fish and other aquatic animals and its salts are frequently found in effluents from industrial wastes. Certain forms of cyanide are acutely toxic to many aquatic life forms and concentrations

<0.1 mg/l can be toxic to some sensitive aquatic species. At the cellular level, cyanide blocks the oxygen consumption of metabolizing cells, which is due to inhibition of the enzyme cytochrome oxidase catalyzing the final oxidation step in cellular respiration. Cyanide forms complexes with some heavy metals such as Zn, Pb, and Cd and is highly toxic. Several cyanide-containing substances display acute toxicity toward aquatic life [38]. However, it has been observed that cyanide-containing compounds also have effects on aquatic life at sublethal concentrations.

4.5 Pharmaceuticals

Pharmaceuticals are among the emerging contaminants in wastewater and are one of the most relevant group of substances having a possible impact on aquatic ecosystems due to their chemicophysical properties [44]. Water bodies that receive wastewater discharges are found to be heavily impacted by annual loadings of these substances. Pharmaceuticals along with their metabolites are readily excreted with urine and feces. While the main concern about pharmaceuticals and their metabolites is that they are being added continuously into lakes and rivers as pollutants, they can have certain adverse effects on aquatic ecosystems and harm freshwater resources including drinking water supplies on a long-term basis. Although concentrations of pharmaceutical compounds in aquatic ecosystems are low, they can cause toxic effects on organisms [45]. Uptake of pharmaceuticals into fish can occur *via* both dermal and gill surfaces for water-borne/sediment-associated pharmaceuticals, orally through the diet, or maternally, *via* the transfer of contaminants through the lipid reserve of eggs. Pharmaceutical drugs are generally designed to have low toxicity but there is the potential for unintended side effects. The active ingredients in pharmaceuticals are known to have potential risks to the aquatic ecosystem and are suspected to have direct toxicity to certain aquatic organisms. There is a global concern about the presence of estrogenic residues in the aquatic ecosystems. The source of these estrogenic residues is industrial wastes and medicines, and as additives in animal feed [46]. The effect of these traces is remarkable on aquatic animals and consequently on humans. Fishes are considered more susceptible to the high concentration of pharmaceuticals. It has been reported that substances such as diclofenac and 17a-ethinylestradiol are responsible for inducing structural disruption in the kidney and intestine and also modify the expression genes, which are associated with the process-controlling metabolism [47, 48]. Their chronic exposure to fishes might affect their survival and reproduction. Another research stated that the presence of antibiotic compounds such as sulfamethoxazole might cause chronic toxicity effects on the photosynthetic apparatus of algae [49]. Therefore, the pharmaceuticals have an effect on the survival of algae due to their rate reduction of photosynthesis by affecting the functions of chloroplasts. A large amount of dead algae lead to secondary effects on the ecosystem such as eutrophication and disruption of the food chain. It threatens the equilibrium of the entire aquatic ecosystem [50].

4.6 Nitrogen

Some water-soluble forms of inorganic nitrogen, such as ionized ammonium, ammonia, nitrite, and nitrate, are present in waste streams, which can exert oxygen demand in surface water resources. Molecular ammonia or NH_4OH is considered as a most toxic form of ammonia, while the dissociated ammonium ion (NH_4) is relatively nontoxic. The discharge of ammonia is mostly from industries, agriculture, and domestic wastewater. Organic wastes contributed from these sources are responsible for the increases in oxygen demand as a result of the increase in biological decomposition and production of ammonia due to the decomposition of organic nitrogen-containing compounds. Ammonia has toxic effects on aquatic life and high concentration can impair aquatic communities [51]. It encourages eutrophication in receiving water bodies. Ammonia and nitrate are principal forms of nitrogen and in the presence of oxygen, ammonia is converted into nitrate creating low dissolved oxygen conditions in surface waters [52, 53]. Excess ammoniacal nitrogen is damaging to aquatic life due to its ability to destroy the aquatic enzyme hydrolysis reaction apart from damaging certain tissues and organs in organisms. Its elevated concentration can cause certain symptoms in aquatic organisms such as hypoxia, coma, and reduced immunity, resulting in slow growth and even large numbers of deaths [54]. Ammonia concentrations >2 mg/l are toxic to aquatic life, especially fishes. Several works done on ammonia toxicity on freshwater vegetation have shown that concentrations >2.4 mg/l inhibit photosynthesis. Further, nitrate causes a decline in amphibian populations and in adverse cases causes poor larval growth, reduced body size, and impaired swimming ability. Direct toxic effects from ammonia are those with a direct impact on individual organisms, typically death, reduced growth rate, or reduced reproductive success.

4.7 Heavy metals

Heavy metals comprise one of the most toxic pollutants in aquatic ecosystems due to the detrimental impacts they display in aquatic biota [55]. The heavy metal present in sewage has severe detrimental effects on the ecological balance of the aquatic environment including organisms [56]. Fishes are among the severely affected species and cannot escape from the detrimental impacts of metals. They accumulate a considerable amount of heavy metals in their body tissues and represent a major dietary source of this element for humans. The presence of heavy metals can inhibit the growth of fish as well as its larvae, reduce the size of fish populations, and can threaten the entire fish population if present in high concentration. A high concentration of aluminum can result in osmoregulatory failure in aquatic animals like fishes [57, 58]. It has the potential to bind with fish gills causing several kinds of diseases, suffocation and ultimately death, change in blood plasma levels, and decrease in nutrient intake at gills. More residence time of water in lakes results in the accumulation of heavy metals in biota, while a significant portion finds its way into the sediments. Mercury has carcinogenic and neuro-toxic properties with the ability to accumulate in living organisms, which gradually increases in the food web. Apart from its toxic effects on humans due to biomagnification in fish, mercury compounds have certain toxic effects on aquatic animals as well.

4.8 Phosphorus

One of the major pollutants found in aquatic environments is phosphorus. The average amount of phosphorus in water resources is <1 mg/l; exceeding the amounts permitted in water causes a serious threat to the environment, animals, and aquatic life. Phosphorous is one of the essential nutrients which promotes algal blooms in rivers and lakes and finally leads to eutrophication which causes oxygen depletion in water *via* algal decay, which has harmful effects on aquatic life. A little rise in the content of this nutrient influences toxin production since it increases the growth of the algae.

5. Eutrophication

5.1 Classification of lakes

Lakes are often classified according to their trophy or degree of enrichment with nutrients and organic matter. They are classified by their trophic state with the main classes of oligotrophic, mesotrophic, eutrophic, and dystrophic (**Table 1**).

Several natural water bodies referred as oligotrophic have clearwater ecosystems with limited primary and secondary productivities due to a shortage of major nutrients [60]. These water bodies under natural succession will require thousands of years to transform into eutrophic. The oligotrophic lake is deep and receiving effluents that are nutrient-poor from its drainage basin. Organic matter production is less in the well-illuminated epilimnion. Therefore, the material sinking into the hypolimnion is the small quantity and little oxygen is consumed there during the summer. In contrast, a eutrophic lake is often, but not necessarily, shallower, the drainage basin is richer, and rivers and groundwater discharge into its epilimnion a substantial amount of nutrients. Primary productivity is higher as compared to that of oligotrophic lakes, and therefore, more organic material settles into the hypolimnion resulting in oxygen depletion. As a result, the deeper layers of water of a eutrophic lake become anoxic during summer. Oligotrophic water bodies have <5–10 μg l^{-1} of phosphorus and < 250–600 μg l^{-1} nitrogen. Oligotrophic water bodies have mean primary productivity ranging between 50 and 300 mg carbon m^{-2} day^{-1}. In eutrophic water bodies, the phosphorus concentration is 10–30 μg l^{-1}, while nitrogen concentration content is 500–100 μg l^{-1}. Primary productivity in eutrophic water bodies is >1 g carbon m^{-2}/day^{-1}. If excessive quantities of phosphorus and nitrogen are added to the water, excessive growth of aquatic plants and

Trophic status	Characteristics	TP ($mg\ m^{-2}$)	TN ($mg\ m^{-2}$)
Oligotrophic	Oligotrophic lakes have poor nutrients and support little plant growth due to which biological productivity is usually low. The waters are clear and the bottom layers have a good oxygen supply throughout the year.	3.0–17.7	307_1630
Mesotrophic	Mesotrophic lakes have transitional characteristics. They are moderately enriched with nutrients and have moderate plant growth.	10.9–95.6	361_1387
Eutrophic	Eutrophic lakes have a rich supply of nutrients that support dense plant growths due to which biological productivity is usually high. The waters are turbid, which support heavy growths of phytoplankton and an abundance of rooted aquatic vegetation. Deepwaters have less concentrations of dissolved oxygen during the seasons of restricted circulation.	16–386	393_6100
Dystrophic	Lakes in the dystrophic state have highly polluted water quality due to which oxygen is absent and the presence of toxins that support no desirable species.	750–1200	—

Table 1.
Lake classification on the basis of trophy or degree of enrichment with nutrients and in relation to P and N [59].

algae takes place. As these algae die, they are decomposed by bacteria and in this process, dissolved oxygen is utilized. The decomposers use up the dissolved oxygen of the water body. Due to this dissolved oxygen, concentrations often fall considerably for fish to breathe resulting in fish kills [61].

5.2 Causes of eutrophication

The term "eutrophic" has been derived from the Greek words eu meaning "well" and trophe meaning "nourishment." Eutrophication refers to the abundant growth of phytoplanktons causing imbalanced primary as well as secondary productivity with a high rate of succession from the existing seral stage to a higher seral stage as a result of nutrient enrichment from fertilizer runoff and humans waste. It takes place at the point when a water body moves toward becoming enriched in key-limiting nutrients, such as nitrates, phosphates, and initiating symptomatic changes, including the expanded production of algae (**Figure 1**). Nitrogen (N) and phosphorus (P) are present in all aquatic ecosystems in some limited amount and are considered as an essential nutrient for the biological growth of organisms. Phosphorus being a macronutrient is essential for all living cells as it is an important constituent of adenosine diphosphate, adenosine triphosphate, nicotinamide adenosine dinucleotide phosphate, nucleic acids as well as phospholipids in the cell wall. Phosphorus is stored as polyphosphates in intracellular volutin granules in prokaryotes as well as eukaryotes. Both N and P are essential nutrients that are required by plants and animals for maintaining their growth and metabolism. However, in wastewater, these essential nutrients are available in abundant as phosphates, combined organic nitrogen, nitrates, and ammonia. On discharge into some receiving water body, their increased concentration can initiate

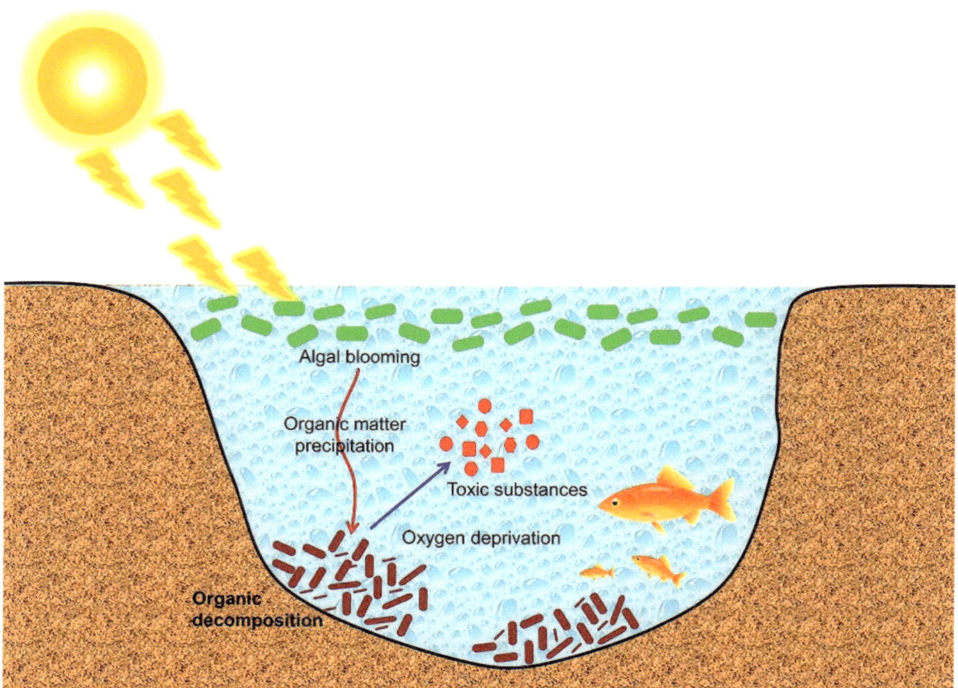

Figure 1.
Eutrophication process [62].

eutrophication with several adverse consequences on the ecological health of the water body [63, 64]. Eutrophication is a natural phenomenon that takes thousands of years to occur in water bodies such as lakes, rivers, and reservoirs. However, an increased rate of nutrient input as a result of anthropogenic activities initiates the process of completing it within a short time period, which is referred to as artificial or cultural eutrophication [65]. Natural eutrophication pushes the succession from open water lake to the marsh to the meadow to the forest, which may take place anywhere within a time period of 500–10,000 years or more depending on the initial condition of that area. Human activities accelerate the rate at which the influx of nutrients into the ecosystems takes place. Runoff resulting from agriculture, urban, and industrial development, mainly from septic systems, sewers, and other human-related actions, increases the rate of entry of both inorganic nutrients and organic substances into aquatic ecosystems.

5.3 Nutrients in aquatic ecosystems

The minimum acceptable concentration of total inorganic phosphate in water is 0.03–0.04 mg l^{-1} and in many lakes, streams, and rivers where the problem of eutrophication is found to occur and its value has been found to increase by 20–25 times during the past 10–15 years especially in cities and industries. Around 60% of the phosphate present in the waterways of the US is contributed from domestic sewage. Phosphate is also contributed from mines, fertilizer runoff, and domestic sewage containing a high concentration of phosphate with about 50% resulting from human waste and 20–30% from detergents. Animal wastes are also rich in nitrate as well as phosphates [66]. Phosphorus resulting from agriculture runoff is the major source of phosphorus loading in riverine sediments, which is being utilized by benthic algae and rooted plants. Eutrophication has become a major concern in many developed as well as developing countries, especially in highly populated countries such as India, China, Bangladesh, Indonesia, and Pakistan. Lakes as well as reservoirs of several industrialized countries of Europe and North America including the Great Lakes of USA and Canada are facing severe threats due to eutrophication. Several lakes of Asia (54%), Europe (53%), North America (48%), South America (41%), and Africa (28%) are eutrophic. As compared to point source pollution, management of diffuse sources is far more challenging due to the difficulty in controlling nutrients contributing from runoff arising from agricultural and urban areas. Most of the phosphorus enters to water body *via* runoff and erosion taking during winter storm events. Thus, phosphorus influx from diffuse sources may be of little significance in the eutrophication of rivers due to the fact that the timing of the transfers does not usually overlap with the period of maximum biological demand. On the other hand, phosphorus being a significant element in the process of eutrophication needs to be identified and quantified from various sources during periods of low flow. Symptoms of eutrophication mostly take place during the plant growing season, that is, spring and summer, when there is a low flow, high water residence times, abundant sunlight light levels, and water temperature is on the higher side, which cause fast algal growth. During the growing period, phosphorus originating from point discharge in rivers is a source of high concentrations of dissolved, bioavailable phosphorus fractions into the water body. According to Meybeck [67], streams and rivers around the world have nearly doubled their concentration of nutrients that is, nitrogen and phosphorus, with local increases of about 50 times. Overall, cultural eutrophication of river ecosystems is a global phenomenon that has, during the past few years, gained much less attention than lake eutrophication. This may be partly due to the effects of increased nutrient concentrations in rivers that are least affected because some factors apart from the nutrients limit algal growth. Although some progress has been made, still there is a less conceptual understanding of eutrophication in rivers and streams. Hydraulic flushing of nutrients, water velocity, and light limitation are indeed significant in regulatory algal growth interacting in several ways. Moreover, short residence time in rivers (<3 days) will have different effects in comparison with longer residence time in impounded rivers or riverine lakes (>3 days). In comparison with lakes (>30 days retention time) and considering some of the factors mentioned above, Hilton [68] devised a conceptual model of how the process of eutrophication takes place in rivers. Since natural streams are net heterotrophic, Dodds [69] formulated the trophic state of rivers into autotrophic, nutrient controlled, and heterotrophic, external carbon-regulated state. The autotrophic state in lotic water bodies is mostly dependent on phosphorus and nitrogen values. Algal biomass is positively correlated with gross primary production in streams and rivers. Eutrophication is a problem that is persistent worldwide. In Spain for example, 80% of the lakes, 70% of the reservoirs, and 60% of the river sites were eutrophic in the 1990s with hypertrophy increasing downstream [70]. There may be several deleterious effects of eutrophication on the environment, which have adverse consequences on the health of the exposed animal population apart from humans through several pathways. Certain health risks appear when extracted freshwater from eutrophic water bodies is supplied for drinking purposes. A severe impact can also occur during animal watering from eutrophic waters.

5.4 Symptoms and effects of eutrophication

The following are the symptoms of eutrophication:

1. Release of limiting nutrients such as phosphorus and nitrogen into the water body.

2. Degradation of water quality such as the appearance of red tides or excessive foam over the surface of the water.

3. Increase in the productivity of the ecosystem along with biomass of phytoplankton, macrophytes, and harmful algal blooms.

4. Reduction in the water clarity and sediments are visible from a depth of few feet. Due to the greenish color of water, turbidity, and high levels of planktonic algae, the clarity of the water is drastically reduced.

5. Oxygen depletion due to increased production of organic matter and formation and release of hydrogen sulfide.

6. Shifts in the composition of species, for example, increased concentration of nitrogen causes new and more competitive forms to invade and compete with original ones.

The following are the effects of eutrophication:

1. Microcystins are certain toxins produced by various genera of cyanobacteria, the predominant one being Microcystis sp. These toxins are highly water-stable and resistant to boiling, and thus pose a threat to water and food quality if not properly monitored. Exposure to microcystins represents a health risk to aquatic organisms, wildlife, domestic animals, and humans upon drinking or

ingesting cyanobacteria in the water. These substances can enter the food chain and cause mortality in an animal apart from other health effects in humans.

2. If the water body affected by eutrophication is used for supplying drinking water to a community, it can cause an increase in the cost of treatment due to prevailing taste and odor problems. Raw water is a source of algae and several other aquatic plants, which also increases the treatment cost, while the quality of water supply may decrease. Planktonic algae when present can shorten filter runs.

3. Certain algae have been found to release organic compounds that are supposed to cause tastes and odors problems besides which they also produce trihalomethanes (THMs) and halo acetic acid (HAA) precursors which are considered as human carcinogens. These compounds react with chlorine, which is used during the disinfection process in wastewater treatment plants and is released with the treated effluent.

4. As the algae die, they become a source of food for the bacterial population, which consumes oxygen during the process. This may cause hypoxia, especially during the night due to which animals especially fish may suffocate resulting in fish kills. Mass deaths are also due to the release of hydrogen sulfide.

5. Aquatic weeds have been often found to block irrigation canals and other water supplies.

6. Excessive growth of macrophytes and algae can impair recreational purposes of water such as swimming, boating, and fishing. Odor problems can arise due to water weeds, dead decaying algae as well as algal scum.

7. Economic loss is also suffered due to change in the composition of species, fish kills, loss of recreational value, and reduction in tourism activities.

6. Conclusion

There has been a continuous increase in the sewage generation throughout the world from domestic, industrial as well as agricultural sources. This has put a serious threat on the freshwater ecosystems as a significant part of them goes untreated into freshwater ecosystems. Due to the presence of a wide variety of contaminants such as suspended solids, pharmaceuticals, heavy metals, sewage disposal has affected several aspects of flora and fauna. It has taken a heavy toll on aquatic life causing several undesirable changes in their structure and composition. Sewage disposal is regarded as a primary culprit in the deterioration of the health of freshwater bodies around the world. It is responsible for the process of eutrophication, which has several negative repercussions on the water bodies including harmful algal blooms, the decline in water quality, loss of economic as well as the esthetic value of the water body.

Conflict of interest

The authors declare no conflict of interest.

Author details

Sami Ullah Bhat* and Umara Qayoom
Department of Environmental Science, School of Earth and Environmental Sciences, University of Kashmir, Srinagar, Jammu and Kashmir, India

*Address all correspondence to: samiullahbhat11@gmail.com

References

[1] Kookana RS, Drechsel P, Jamwal P, Vanderzalm J. Urbanisation and emerging economies: Issues and potential solutions for water and food security. Science of the Total Environment. 2020;**732**:139057. DOI: 10.1016/j.scitotenv.2020.139057

[2] Beiras R, Bellas J, Cachot J, Cormier B, Cousin X, Engwall M, et al. Ingestion and contact with polyethylene microplastics does not cause acute toxicity on marine zooplankton. Journal of Hazardous Materials. 2018;**360**: 452-460. DOI: 10.1016/j.jhazmat.2018.07.101

[3] Mateo-Sagasta J, Zadeh SM, Turral H, Burke J. Water pollution from agriculture: A global review. Executive Summary. Food and Agriculture Organization of the United Nations Rome and International Water Management Institute, 2017

[4] Vörösmarty CJ, McIntyre PB, Gessner MO, Dudgeon D, Prusevich A, Green P, et al. Global threats to human water security and river biodiversity. Nature. 2010;**467**(7315):555-561. DOI: 10.1038/nature09440

[5] Qayoom U, Bhat SU, Ahmad I. Efficiency evaluation of sewage treatment technologies: Implications on aquatic ecosystem health. Journal of Water and Health. 2020;**19**(1):29-46

[6] Mhlanga L, Day J, Cronberg G, Chimbari M, Siziba N, Annadotter H. Cyanobacteria and cyanotoxins in the source water from Lake Chivero, Harare, Zimbabwe, and the presence of cyanotoxins in drinking water. African Journal of Aquatic Science. 2006;**31**(2): 165-173. DOI: 10.2989/16085910609503888

[7] Havens K.E. Cyanobacteria blooms: effects on aquatic ecosystems. In: Hudnell H.K. (eds) Cyanobacterial Harmful Algal Blooms: State of the Science and Research Needs. Advances in Experimental Medicine and Biology. New York, NY: Springer; 2008. vol 619. https://doi.org/10.1007/978-0-387-75865-7_33

[8] Zhang X, Yang L, Li Y, Li H, Wang W, Ye B. Impacts of lead/zinc mining and smelting on the environment and human health in China. Environmental Monitoring and Assessment. 2012;**184**(4):2261-2273. DOI: 10.1007/s10661-011-2115-6

[9] Raschid-Sally L, Jayakody P. Drivers and Characteristics of Wastewater Agriculture in Developing Countries: Results from a Global Assessment. Colombo, Sri Lanka: International Water Management Institute; 2009

[10] Yu S, Miao C, Song H, Huang Y, Chen W, He X. Efficiency of nitrogen and phosphorus removal by six macrophytes from eutrophic water. International Journal of Phytore mediation. 2019;**21**(7):643-651. DOI: 10.1080/15226514.2018.1556582

[11] Ajonina C, Buzie C, Rubiandini RH, Otterpohl R. Microbial pathogens in wastewater treatment plants (WWTP) in Hamburg. Journal of Toxicology and Environmental Health, Part A. 2015;**78**(6):381-387. DOI: 10.1080/15287394.2014.989626

[12] Al-Gheethi AA, Efaq AN, Bala JD, Norli I, Abdel-Monem MO, Kadir MA. Removal of pathogenic bacteria from sewage-treated effluent and biosolids for agricultural purposes. Applied Water Science. 2018;**8**(2):1-25. DOI: 10.1007/s13201-018-0698-6

[13] Grandclément C, Seyssiecq I, Piram A, Wong-Wah-Chung P, Vanot G, Tiliacos N, et al. From the conventional biological wastewater treatment to hybrid processes, the evaluation of

organic micropollutant removal: A review. Water Research. 2017;**111**:297-317. DOI: 10.1016/j.watres.2017.01.005

[14] Osuolale O, Okoh A. Human enteric bacteria and viruses in five wastewater treatment plants in the Eastern Cape, South Africa. Journal of Infection and Public Health. 2017;**10**(5):541-547. DOI: 10.1016/j.jiph.2016.11.012

[15] Kirchmann H, Pettersson S. Human urine-chemical composition and fertilizer use efficiency. Fertilizer Research. 1994;**40**(2):149-154. DOI: 10.1007/BF00750100

[16] Larsen TA, Alder AC, Eggen RI, Maurer M, Lienert J. Source separation: Will we see a paradigm shift in wastewater handling? Environ. Sci. Technol. 2009, 43, 6121-6125. DOI: 10.1021/ es803001r

[17] World Waterfall Database. (2017). Niagara Falls, Ontario, Canada. Retrieved from: https://www.worldwaterfalldatabase.com/index.php/waterfall/Niagara–Falls–106

[18] Qadri R, Faiq MA. Freshwater pollution: Effects on aquatic life and human health. In: Fresh water pollution dynamics and remediation. Singapore: Springer; 2020. pp. 15-26. DOI: 10.1007/978-981-13-8277-2_2

[19] World Water Assessment Programme (UNESCO WWAP) 2017. Available from: http://www.unesco.org/new/en/natural-sciences/environment/water/wwap

[20] Sato T, Qadir M, Yamamoto S, Endo T, Zahoor A. Global, regional, and country level need for data on wastewater generation, treatment, and use. Agricultural Water Management. 2013;**130**:1-3. DOI: 10.1016/j.agwat.2013.08.007

[21] Reckson K, Tawanda MS. Effect of sewage disposal on the water quality in Marimba River and Lake Chivero: The case of Crowborough grazing land of Harare. Zimbabwe. International Researcher. 2012;**1**(3):53-57

[22] Horner RR, Skupien JJ, Livingstone EH and Shaver HE. Fundamentals of Urban Runoff Management: Technical and Institutional Issues. Prepared by the Terrene Institute, Washington, DC, in cooperation with the U.S. Environmental Protection Agency. 1994. EPA/840/B-92/002

[23] Gupta P, Vishwakarma M, Rawtani PM. Assessment of water quality parameters of Kerwa Dam for drinking suitability. International Journal of Theoretical and Applied Sciences. 2009;**1**(2):53-55

[24] Lecerf A, Risnoveanu G, Popescu C, Gessner MO, Chauvet E. Decomposition of diverse litter mixtures in streams. Ecology. 2007;**88**(1):219-227

[25] Baerlocher F, Seena S, Wilson KP, Dudley WD. Raised water temperature lowers diversity of hyporheic aquatic hyphomycetes. Freshwater Biology. Feb 2008;**53**(2):368-379

[26] Dojlido J, Best GA. Chemistry of Water and Water Pollution. Ellis Horwood Limited; 1993

[27] Mayer FL Jr, Ellersieck MR. Experiences with single-species tests for acute toxic effects on freshwater animals. Ambio. 1988;**17**:367-375

[28] Whitehead PG, Hornberger GM. Modelling algal behaviour in the River Thames. Water Research. 1984;**18**(8): 945-953

[29] Wade AJ, Whitehead PG, Hornberger GM, Snook DL. On modelling the flow controls on macrophyte and epiphyte dynamics in a lowland permeable catchment: The River Kennet, southern England.

Science of the Total Environment. 2002;**282**:375-393

[30] Hawkins CP, Hogue JN, Decker LM, Feminella JW. Channel morphology, water temperature, and assemblage structure of stream insects. Journal of the North American Benthological Society. 1997;**16**(4):728-749

[31] Daufresne M, Roger MC, Capra H, Lamouroux N. Long-term changes within the invertebrate and fish communities of the Upper Rhône River: Effects of climatic factors. Global Change Biology. 2004;**10**(1):124-140

[32] Connolly NM, Crossland MR, Pearson RG. Effect of low dissolved oxygen on survival, emergence, and drift of tropical stream macro invertebrates. Journal of the North American Benthological Society. 2004;**23**(2):251-270

[33] Olabode GS, OF O, Somerset VS. Physicochemical properties of wastewater effluent from two selected wastewater treatment plants (Cape Town) for water quality improvement. International journal of Environmental Science and Technology. 2020;**17**: 4745-4758. DOI: 10.1007/s13762-020-02788-9

[34] Suthar S, Sharma J, Chabukdhara M, Nema AK. Water quality assessment of river Hindon at Ghaziabad, India: Impact of industrial and urban wastewater. Environmental Monitoring and Assessment. 2010;**165**(1):103-112. DOI: 10.1007/s10661-009-0930-9

[35] Nelsen TA, Blackwelder P, Hood T, McKee B, Romer N, Alvarez-Zarikian C, et al. Time-based correlation of biogenic, lithogenic and authigenic sediment components with anthropogenic inputs in the Gulf of Mexico NECOP study area. Estuaries. 1994;**17**(4):873-885. DOI: 10.2307/1352755

[36] Frodge JD, Thamas GL, Pauley GB. Effects of canopy formation by floating and submergent aquatic macrophytes on the water quality of two shallow Pacific Northwest lakes. Aquatic Botany. 1990;**38**:231-248

[37] Lovley DR, Giovannoni SJ, White DC, Champine JE, Phillips EJ, Gorby YA, et al. Geobacter metallireducens gen. nov. sp. nov., a microorganism capable of coupling the complete oxidation of organic compounds to the reduction of iron and other metals. Archives of Microbiology. 1993;**159**(4):336-344. DOI: 10.1007/BF00290916

[38] Brett JR, Blackburn JM. Oxygen requirements for growth of young coho (Oncorhynchus kisutch) and sockeye (O. nerka) salmon at 15 C. Canadian Journal of Fisheries and Aquatic Sciences. 1981;**38**(4):399-404. DOI: 10.1139/f81-056

[39] Chambers PA, Mill TA. Dissolved oxygen conditions and fish requirements in the Athabasca, Peace and Slave Rivers: Assessment of present conditions and future trends. In: Northern River Basins Study, Edmonton, Alberta; 1996

[40] Lake RG, Hinch SG. Acute effects of suspended sediment angularity on juvenile coho salmon (Oncorhynchus kisutch). Canadian Journal of Fisheries and Aquatic Sciences. 1999;**56**(5):862-867. DOI: 10.1139/f99-024

[41] Van Nieuwenhuyse EE, LaPerriere JD. Effects of placer gold mining on primary production in subarctic streams of alaska 1. Jawra Journal of the American Water Resources Association. 1986;**22**(1):91-99. DOI: 10.1111/j.1752-1688.1986.tb01864.x

[42] Birkett C, Tollner EW, Gattie DK. Total suspended solids and flow regime

effects on periphyton development in a laboratory channel. Transactions of the ASABE. 2007;**50**(3):1095-1104

[43] Slaney PA, Halsey TG, Tautz AF. Effects of forest harvesting practices on spawning habitat of stream salmonids in the Centennial Creek watershed, British Columbia. Ministry of Recreation and Conservation, Victoria, B.C. Fish Manage Rep; 1997;**73**:45

[44] Bottoni P, Caroli S, Caracciolo AB. Pharmaceuticals as priority water contaminants. Toxicological and Environmental Chemistry. 2010;**92**(3): 549-565. DOI: 10.1080/0277224100 3614320

[45] Peñuela GA, Martínez-López E. Enzymatic activity changes in striped catfish Pseudoplatystoma magdaleniatum, induced by exposure to different concentrations of ibuprofen and triclosan. Chemosphere. 2021;**271**:129399. DOI: 10.1016/j. chemosphere.2020.129399

[46] Seelig B. Water Resource Impacts from Medicines and Other Biologically Active Substances. Available from: http://www.ag.ndsu.edu/pubs/h2oqual/ watgrnd/wq1278.pdf

[47] Mehinto AC, Hill EM, Tyler CR. Uptake and biological effects of environmentally relevant concentrations of the nonsteroidal anti-inflammatory pharmaceutical diclofenac in rainbow trout (Oncorhynchus mykiss). Environmental Science and Technology. 2010;**44**(6): 2176-2182

[48] Lyssimachou A, Arukwe A. Alteration of brain and interrenal StAR protein, P450 scc, and Cyp11β mRNA levels in atlantic salmon after nominal waterborne exposure to the synthetic pharmaceutical estrogen ethynylestradiol. Journal of Toxicology and Environmental Health, Part A. 2007;**70**(7):606-613

[49] Liu BY, Nie XP, Liu WQ, Snoeijs P, Guan C, Tsui MT. Toxic effects of erythromycin, ciprofloxacin and sulfamethoxazole on photosynthetic apparatus in Selenastrum capricornutum. Ecotoxicology and Environmental Safety. 2011;**74**(4): 1027-1035

[50] Lanzky PF, Halting-Sørensen B. The toxic effect of the antibiotic metronidazole on aquatic organisms. Chemosphere. 1997;**35**(11):2553-2561

[51] Versteeg DJ, Belanger SE, Carr GJ. Understanding single-species and model ecosystem sensitivity: Data-based comparison. Environmental Toxicology and Chemistry: An International Journal. 1999;**18**(6):1329-1346. DOI: 10.1002/etc.5620180636

[52] Sabalowsky AR. An investigation of the feasibility of nitrification and denitrification of a complex industrial wastewater with high seasonal temperatures [Doctoral dissertation]. Virginia Tech

[53] Kurosu O (2001) Nitrogen removal from wastewaters in microalgalbacterial-treatment ponds. Available from: http://www.socrates. berkeley.edu/es196/projects/2001final/ kurosu.pdf

[54] Krakat N, Demirel B, Anjum R, Dietz D. Methods of ammonia removal in anaerobic digestion: A review. Water Science and Technology. 2017;**76**(8): 1925-1938. DOI: 10.2166/wst.2017.406

[55] Qayoom U, Bhat SU, Ahmad I, Kumar A. Assessment of potential risks of heavy metals from wastewater treatment plants of Srinagar city, Kashmir. International journal of Environmental Science and Technology. 2021;**30**:1-20

[56] Farombi EO, Adelowo OA, Ajimoko YR. Biomarkers of oxidative stress and heavy metal levels as

indicators of environmental pollution in African cat fish (Clarias gariepinus) from Nigeria Ogun River. International Journal of Environmental Research and Public Health. 2007;**4**(2):158-165. DOI: 10.3390/ijerph2007040011

[57] Vosylienė MZ, Mikalajūnė A. Effect of heavy metal model mixture on rainbow trout biological parameters. Ekologija. 2006;**4**:12-17

[58] Rosseland BO, Eldhuset TD, Staurnes MJ. Environmental effects of aluminium. Environmental Geochemistry and Health. 1990;**12**(1): 17-27

[59] Salameh E, Harahsheh S. Eutrophication processes in arid climates. In: Eutrophication: Causes, Consequences and Control. Dordrecht: Springer; 2010. pp. 69-90

[60] Beeby A. What do sentinels stand for? Environmental Pollution. 2001; **112**(2):285-298

[61] Murphy KJ. Plant communities and plant diversity in softwater lakes of northern Europe. Aquatic Botany. 2002;**73**(4):287-324. DOI: 10.1016/S0304-3770(02)00028-1

[62] Umbria A. Stato di qualità ambientale del laghi e analisi dei trend evolutivi. In: Documento Tecnico. 2009. Available from: http://www.arpa.umbria.it/resources/documenti/Acqua/Rapporto%20laghi%202005-2006-2007.pdf

[63] Khan FA, Ansari AA. Eutrophication: An ecological vision. The Botanical Review. 2005;**71**(4): 449-482

[64] Fink G, Alcamo J, Flörke M, Reder K. Phosphorus loadings to the world's largest lakes: Sources and trends. Global Biogeochemical Cycles. 2018;**32**(4):617-634. DOI: 10.1002/2017GB005858

[65] Rovira JL, Pardo P. Nutrient pollution of waters: Eutrophication trends in European marine and coastal environments. Contributions to Science. 2006:181-186. DOI: 10.2436/20.7010.01.4

[66] Penelope RV, Charles RV. Water resources and the quality of natural waters. London: Jones and Bartbett Publishers; 1992

[67] Meybeck M. Carbon, nitrogen, and phosphorus transport by world rivers. American Journal of Science. 1982; **282**(4):401-450

[68] Hilton J, O'Hare M, Bowes MJ, Jones JI. How green is my river? A new paradigm of eutrophication in rivers. Science of the Total Environment. 2006;**365**(1-3):66-83. DOI: 10.1016/j.scitotenv.2006.02.055

[69] Dodds WK. Eutrophication and trophic state in rivers and streams. Limnology and Oceanography. 2006;**51**(1part2):671-680

[70] Cobelas MA, Jacobsen BA. Hypertrophic phytoplankton: An overview. Freshwater Forum. 1992;**2**(3):184-199

Chapter 7

Biological versus Physicochemical Technologies for Industrial Sewage Treatment: Which Is the Most Efficient and Inexpensive?

Karima Elkarrach, Fatima Atia, Anass Omor, Omar Laidi, Saloua Biyada, Mohamed Benlmelih and Mohammed Merzouki

Abstract

Industries play a major role in the development of countries' economy. However, they are known as the biggest source of water pollution in the whole world. In fact, several industries use a huge amount of water in their manufacturing operations, and then, they reject a large volume of wastewaters such as tanneries, brassware, olive mills ... etc. The sewage of these industries may contain organic/inorganic matters or toxic components that harm human health and the environment. Therefore, the treatment of these effluents is necessary. For that, there are many treatment processes, including biological and physicochemical processes or both. The choice of adequate process is depending on many reasons, especially on the biodegradability degree of each effluent, as well as the presence of recalcitrant pollutants. Nevertheless, biological technologies, particularly bioremediation, are recently an emerging technology for the elimination of recalcitrant pollutants like heavy metals. Furthermore, these biotechnologies are simple, efficient, eco-friendly and inexpensive. Therefore, this environmental biotechnology may be a new approach for the treatment of industrial sewage, so, it can successfully replace physicochemical technologies that are very expensive.

Keywords: Industries, sewage, biological technologies, physicochemical processes

1. Introduction

Water is so essential in our daily life. However, water resources and their quality are progressively decreasing because of human activities, and then several countries are threatened by water scarcity, including Morocco. So, the rational management of these water resources is a big challenge in the whole world. Nowadays, the treatment and reuse of wastewater are known as the best solutions to deal with this lack of water. The constraint of this issue is pollution, especially industrial pollution, because industries generate high toxic substances, which may reduce the performance of the treatment [1]. For example in Morocco, the large hydraulic basin 'Sebou' receives more than 40% of pollution from industries of Fez city such as tanneries, textiles, brassware, olive mill ... etc.

Industries release organic and inorganic pollutants namely heavy metals, dyes, polyphenols ... etc. Heavy metals are toxic, in which their toxicity depends on several factors, particularly the metal dose and the time exposition. Arsenic, cadmium, chromium, lead, and mercury, have ranked as carcinogenic metals, because they may damage multiple organs even at lower exposure doses [2]. Likewise, textile dyes are highly toxic and potentially carcinogenic [3]. Thus, they can lead to various animal and human diseases, as well as the environmental degradation. Therefore, recalcitrant substances harm the environment and human health, and then their removal from wastewaters is mandatory.

The literature has shown several physicochemical and biological processes for the treatment of industrial sewage. Among physicochemical processes, there is coagulation [4], electrocoagulation [5], forward osmosis [6], chemical precipitation [7], adsorption [8], and oxidation [9]. As for biological systems, there are many technologies such as sequencing batch reactor [10], bioaugmentation [11], biosorption [12], membrane bioreactor [13], anaerobic digestion [14]... etc. In fact, each process has its advantages and disadvantages; hence the process performance is highly dependent on the nature of the effluent and the flow to be treated. For example, physical–chemical treatments are known for their high performance, but they are very expensive and can generate another serious pollution. Biological treatments are also efficient and ecological, but the presence of recalcitrant substances in huge amount can decrease their efficient. So, the issue is complex, because it is necessary to find a treatment process that will be eco-friendly, efficient, and economical at the same time.

Taking into account the above, this chapter focuses on various physicochemical and biological processes for the treatment of industrial sewage. Moreover, this chapter will show the effectiveness of biological technologies for the removal of toxic substances.

2. Treatment of industrial wastewaters

The treatment of industrial effluents is essential before their discharge into the natural environment. Several physicochemical and biological treatments of these effluents have been studied in the literature. These techniques have been considered simple, efficient, or even advanced, but each system has certain advantages over the other. Moreover, these treatment systems can be applied independently or combined.

2.1 Physicochemical treatments

2.1.1 Membrane filtration

2.1.1.1 Reverse osmosis

This technology is based on the use of a semi-permeable membrane, wherein pollutants will be captured. This treatment system is known for its high purification rate, and it may be used for the treatment of all industrial sewage [15]. A study has shown that reverse osmosis is an advanced and promising technique for industrial wastewaters. Despite the qualities of this treatment system, it presents certain disadvantages, particularly the high cost.

2.1.1.2 Membrane filtration

Currently, this technique is well developed; it is based on the physical separation of pollutants under hydraulic pressure. This treatment system is very efficient for industrial sewage treatment such as pharmaceutical [16], textile [17], pulp and paper effluents [18]...etc. The separation is based on three principles, which are adsorption, electrostatic phenomenon, and sieving [19].

According to the membrane's pore size, there are three filtration types:

- Microfiltration: it refers to remove substances with a size greater than 10 μm through a membrane with pores between 0.1–10 μm. It is characterized by the tangential passage, and it is done under low pressure gradient of 1–3 bars. This technique is effective for sewage purification and the elimination of microorganisms, especially bacteria.

- Ultrafiltration: the pores of the membrane are between 0.001 and 0.1 μm. Thanks to these small pores, only water and small molecules (Ions) can pass through this membrane, while macromolecules such as proteins, polymers, bacteria, and viruses, will be retained [20].

- Nanofiltration: the pore size of the membrane is less than 1 nm. This technique is effective for industrial sewage treatment. It may remove heavy metals and ions as chromium and nickel [21]. It can also retain organic substances with a molecular weight fewer than 300 daltons. Nevertheless, it could not produce demineralized water as the reverse osmosis technique.

Otherwise, the membrane can be mineral (metallic, ceramic, etc.) or organic (polyamides, cellulose acetate, etc.). Its structure can be uniform (Isotropic) or composite (Anisotropic). Indeed, organic membranes are the most used because of their low cost, but mineral membranes can resist extreme conditions (Temperature, pH, etc.).

Consequently, membrane filtration has several qualities, namely the removal of micro-organisms, heavy metals, turbidity, dyes, and also odors from industrial sewage. Despite these advantages, the technique has also some limits such as rapid membrane fouling, production of high amount of sludge, high investment costs, and high energy consumption.

2.1.2 Coagulation-flocculation

The coagulation-flocculation process involves the use of coagulant and flocculant agents that can regroup the pollutants together as heavy flocs. These flocs will be eliminated by precipitation or filtration. These agents may be iron or aluminum chemicals. According to Junio et al. [4], coagulation-flocculation is a simple, fast, and effective technique for removing pollutants from industrial wastewaters. In this study, ferric chloride was used as a coagulant agent for the treatment of tannery sewage, wherein the abatement rates of COD and suspended solids were above 80%. Nevertheless, this technique has several disadvantages, namely the production of high sludge and the increase of acidity and conductivity within the treated effluent. On the other hand, the use of bio-coagulants and bio-flocculants is a new approach of this technique in order to reduce the massive use of chemicals and their harmful effects. According to previous study, cactus juice can be used as a bio-flocculant to reduce chromium, in which chromium VI removal was around 98% [22].

2.1.3 Electrocoagulation

Electrocoagulation has considered as a new alternative of chemical coagulation, and it is a promising process for the treatment of industrial sewage [23]. This technique is based on the principle of soluble anodes, and it induces the electrochemical separation of pollutants. These anodes are often made of aluminum or iron, from where metal cations (Fe^{3+} or Al^{3+}) are generated by imposing an electric current between these anodes. These metal cations react as a coagulant to destabilize the suspended particles, and then, the formation of flocs that will subsequently precipitate. Indeed, this technique has several advantages, but the production of sludge in high amount and the consumption of high energy are its main disadvantages.

2.1.4 Oxidation

Oxidation is based on electrochemical reactions between the oxidizing agent and the pollutant by changing the electrons. This technique aims to modify the characterization of refractory pollutants by making them insoluble to facilitate their elimination, or soluble but non-toxic. The most commonly used chemical oxidants are oxygen, hydrogen peroxide, chlorine, ozone, potassium permanganate, and ferric chloride. This system is known for its strong elimination of bad odors, either natural or produced during anaerobic conditions. In addition, the combination of ozone with ultraviolet rays (UV) or hydrogen peroxide produces free radicals that are powerful oxidants and can eliminate a large part of the COD [24]. The advantages and disadvantages of this process are depending on the oxidant agent and the type of pollutant (**Table 1**).

According to the literature, several types of research have shown the efficiency of this process for the elimination of sulfides, of which hydrogen peroxide is the most used. This oxidant can eliminate 85–100% of sulfides by using 1.3 to 4.0 mg/L of H_2O_2 for 1 mg/L of sulfides [26].

In recent decades, this process has been developed using the combination of two powerful chemical oxidants (H_2O_2/Fe^{2+} and H_2O_2/O_3), photo-catalysis (UV),

Oxidant	Advantages	Disadvantages
Oxygen	Low investment costs. Simple process	Incomplete oxidation. Production of colloidal sulfur and poly-sulfides. Difficult to control and build. Increase of the turbidity.
Chlorine	Low investment costs. Simple process	Incomplete oxidation. Use of high dose. Lack of security. Increase the turbidity
Ozone	Easy process Production of high water quality	Expensive process Increase the turbidity Ozone concentration is low than 2 mg/L
Potassium Permanganate	Easy and economical process	Requires the use of filters for the removal of residual MnO_2. Increase the turbidity. Use of a large quantity of the product. Expensive product.
Hydrogen peroxide	Easy and economical process	Incomplete oxidation. Increase the turbidity Use of high amount of products. Requires a long contact time.
Ferric chloride	Economical and powerful oxidant	This process has not been demonstrated on a pilot scale for the treatment of drinking water.

Table 1.
Advantages and disadvantages of oxidation agents [25].

sonochemical oxidation, or electrochemical oxidation. As a result, this oxidation is called advanced oxidation (POA), which is considered an innovative process and an emerging technology for the treatment of industrial sewage. The principle of advanced oxidation is based on the production of hydroxyl radicals, which are very active and react rapidly on organic and inorganic compounds [9]. This advanced process has several qualities like high removal rates in a short time and minimal sludge production. However, it also has some disadvantages namely the high operating and investment costs, as well as the treatment efficiency depends on nature and pollutant concentration.

2.1.5 Adsorption

Adsorption is well known for the treatment of industrial sewage, so it is characterized by its high purifying capacity [27]. This process is based on the use of a material that will retain pollutants. This material can be applied as a fixed bed or can be used in suspension with the effluent (Fluidized Bed). The adsorption material can be inorganic (rocks, ashes ...), or organic (fruit, vegetables, wood, bacteria ....). Moreover, it can be used natural or activated.

The activation of a material can be carried out with a physical process (Pyrolysis, calcination, carbonization), a chemical process (Acids, bases), or the both. Physical activation involves high temperatures (800–1000°C), whereas chemical activation requires low temperatures. The most commonly used chemical agents are potassium hydroxide, sodium bicarbonate, sodium hydroxide, zinc chloride, sulfuric and phosphoric acids. However, phosphoric acid is frequently used comparing to other agents because of its low cost and activation temperature, as well as this acid can be recoverable (< 600°C) [27]. On the other hand, the activation of materials with potassium hydroxide is considered as the most effective [28]. In fact, chemical activation has several advantages, especially the increase in the specific exchange surface and the material porosity [8]. Although the activated carbon is a powerful adsorbent, its use is so limited due to its high cost. For that, this technique will be very attractive and promising if the material will be efficient and inexpensive at the same time. Thus, several attempts have been made to find novel materials such as organic waste, fly ash [29], olive pomace [30], and sawdust [31].

2.1.5.1 Adsorption types

According to the literature, the principle of adsorption is based on the binding forces between ions of adsorbent and adsorbate. Consequently, the involved forces divide adsorption into two types:

- Physical adsorption or physisorption: this type of adsorption is characterized by the absence of electron exchange between the adsorbent and the adsorbate. However, the adsorbate is retained by the adsorbent through non-specific physical forces of Van der Waals type, where multiple layers can be formed. This adsorption type requires low heat and it is reversible and non-specific.

- Chemical adsorption or chemisorptions: it is based on the exchange of electrons between the adsorbent and the adsorbate. This process requires high energy compared to physical adsorption. This energy corresponds to the eternal covalent bonds between ions of the adsorbent and the adsorbate, and then the phenomena of ion exchange and protonation/deprotonation are the main mechanisms. Moreover, a single layer could be only formed through this adsorption type, while other layers can be retained by physisorption.

2.1.5.2 Adsorption isotherms and kinetics

The study of isotherms is essential because it expresses the static adsorption capacity for an adsorbent/adsorbate couple. There are four main types of isotherms, which are as follows (**Figure 1**):

- Type a, which reveals cooperative adsorption of the adsorbate molecules.
- Type b or "Langmuir type", which is observed in the case of progressive microporous adsorption.
- Type c, which is observed in the case of a strong interaction between the adsorbate and the surface of the solid.
- Type d or linear isotherm, which occurs when the solutes penetrate more easily into the solid than into the solvent [32].

On the other hand, several mathematical models describe the adsorption mechanisms, where Langmuir, Freundlich, and Temkin are widely used.

- Langmuir's model: indicates that the adsorption is monolayer and into homogeneous surface. It is the oldest and the most common model. A limited adsorption capacity (qmax) is retained by the solid.
- Freundlich model: this empirical model reveals that the material surface is heterogeneous.
- Temkin's model: It takes into account adsorbant-adsorbate interactions.

Adsorption kinetics allows also a better understanding of the adsorption mechanism. As well, it describes the adsorption rate that leads to the control of the equilibrium time. The mathematical models of the adsorption kinetics are numerous and they can be used for the optimization of treatment models. Among these models, we quote pseudo-first-order, pseudo-second-order, and Elovich.

2.2 Biological treatments

Although the COD/BOD_5 ratio reveals that physicochemical treatments are the most suitable for inorganic industrial sewage, the literature has shown that

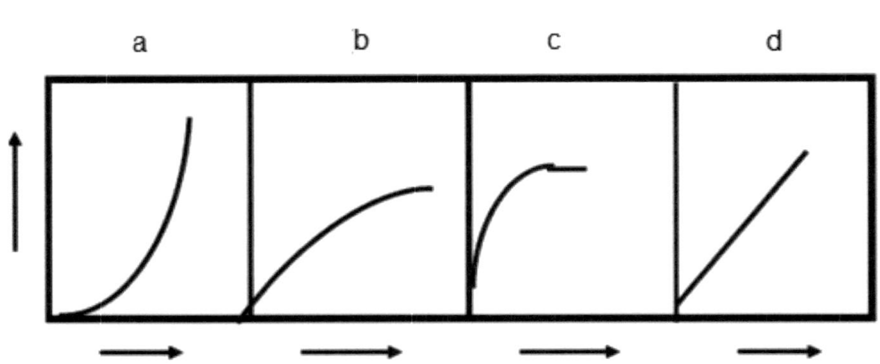

Figure 1.
Main types of adsorption isotherms according to [32].

biological treatments are also efficient and promising for the treatment of these effluents. Biological treatment involves the use of non-specific microorganisms (Activated sludge), well-selected microorganisms (Bio-augmentation), algae, or plants (Phytoremediation). These microorganisms can be bacteria, yeasts, fungi, etc.

Biological treatment consists of the use of pollutants of industrial sewage as a source of nutrition and energy by these microorganisms for their growth. This type of treatment has several advantages comparing to physicochemical treatments. Among these qualities, there are:

- Easy installation and control on a large scale.
- Attractive and economical processes.
- Ecological processes.
- Very effective for the removal of biodegradable organic matter, nitrogen, and phosphorus in particular.

Furthermore, the presence or absence of oxygen divides these biological treatments into two categories, which are aerobic and anaerobic treatments.

2.2.1 Aerobic treatments

2.2.1.1 Classical process of activated sludge

The activated sludge process is the most classic and most famous process since the 20th century. It is based on the treatment with aerobic micro-organisms. These micro-organisms are generally autotrophic or heterotrophic, and they are composed of several groups like bacteria (Gram-positive and/or Gram-negative), fungi, yeasts, protozoa, and metazoa. Consequently, activated sludge is composed of micro-organisms plus inert, organic or mineral matters. These elements are grouped through mucilaginous substances. In fact, those micro-organisms degrade the organic matter of wastewaters into carbon dioxide, water, new cells, and other non-toxic by-products. The separation between the treated effluent and the activated sludge takes place in a clarifier or decanter.

Even this process is very effective in the degradation of organic sewage like agri-food sewage; it also has some disadvantages, particularly the bulking phenomenon, the presence of high amount of recalcitrant pollutants, as well as nitrogen and phosphorus are not reduced at the same time and in the same reactor. This requires the addition of another tank to remove them, and subsequently the increase of installation and operation costs.

2.2.1.2 Sequencing batch reactor

The Sequencing Batch Reactor (SBR) is an emerging approach that has been well developed recently. This technique has the same treatment principle as the previous process, so it is based on the treatment by activated sludge under aerobic conditions. However, this process differs from the other one, because all treatment steps are performed in a single reactor, including nitrogen and phosphorus removal. In addition, the separation liquid/solid is done in the same reactor, where the good separation is linked to the presence of an adequate quantity of filamentous bacteria.

The treatment through this system is by cycle that involves four successive phases, which are:

- Phase one is the supply of the reactor with the effluent to be treated.
- Phase two is the aeration or treatment phase. In this phase, the effluent is brought into contact with the activated sludge in the presence of oxygen. During this stage, the effluent will be degraded by micro-organisms.
- Phase three is the settling phase during which the treated effluent will be separated from the activated sludge.
- Phase four is the withdrawal phase, from which the treated effluent will be withdrawn and the new cycle will be started.

The performance of this system depends on many parameters such as the volumetric organic load, daily cycle number of the treatment, level of dissolved oxygen, sludge index, sludge age, and time of the settling phase. The optimization of these parameters leads to very high abatement rates.

The reference [33] showed that SBR is a promising technique for the treatment of several types of sewage such as agri-food, pharmaceutical, pulp/paper, textile, tannery, chemical, and petrochemical effluents, ... etc.

Another study shows the efficiency of the SBR system compared to the conventional activated sludge process (**Table 2**). So, the SBR is the most efficient due to very high removal rates, and also the system is capable to remove nitrogen and phosphorus at the same time. Thus, this system is inexpensive and useful than the classic process. Furthermore, aerobic denitrification was highlighted within this reactor [10, 35]. In recent decades, this phenomenon has been very attractive and advanced according to various studies. The literature showed more than 37 aerobic denitrifying bacteria, where *Bacillus pulminus*, *Arthrobacter sp.*, and *Streptomyces lusitanus* were the latest shown [36].

Heavy metals could also be removed by this biological system. The reference [11] has shown that this system is capable to remove chromium from a tannery effluent with a removal rate of 96.1% using a low volumetric organic load. As well, [37] has indicated that several heavy metals (Nickel, chromium, cadmium, cobalt, zinc, and silver) were eliminated from brassware effluent through this system with high removal rates that reached more than 60%.

This system has many advantages such as low cost, short treatment time compared to the classic process, high removal of organic and mineral matter, simplicity of the process, the possibility of spreading excess sludge, limitation of bad odors,

Parameters	Abatement rate (%)	
	Sequencing batch reactor	Classic process of activated sludge
BOD_5	89–99	85–95
Suspended solids	85–97	85–90
Total nitrogen	>75	Untreated
Phosphorus	57–69	Untreated
Total coliforms	99	90–96

Table 2.
Average abatement rates obtained by the sequential batch reactor and the conventional activated sludge process [34].

etc. ... Despite all these advantages, it also has some shortcomings like the bulking phenomenon.

2.2.1.3 Membrane bioreactor

The membrane bioreactor is a new alternative to the classic activated sludge process, so it is based on the same principle of activated sludge treatment. However, the solid/liquid separation is done through membrane column instead of clarifier. Consequently, this technique is the combination of an activated sludge biological reactor and a membrane process such as microfiltration. It is widely used for the treatment of industrial sewage [13]. Furthermore, the use of membrane filtration increases the rate of effluent purification due to its removing capacity of high concentrations of suspended solids, nitrogen and phosphorus, as well as bacteria and viruses. However, membrane cleaning or regeneration after plugging is essential, so it increases considerably the process cost.

2.2.1.4 Bioremediation

The presence of a high concentration of heavy metals, salts, or other toxic substances, reduces or prevents the treatment by the activated sludge because of these extreme conditions. This issue allows us to highlight biotechnological technique that is bioremediation. This biotechnology regroups some processes like bio-augmentation, biosorption and phytoremediation. These techniques use a powerful microorganism, consortium, or plant, which can resist these extreme conditions.

Bioaugmentation is based only on the use of living microorganisms, whereas biosorption involves living or non-living microorganisms. Moreover, biosorption is one of the various mechanisms of bioaugmentation. For biosorption, the microorganisms can replace the activated carbon, and then reduce process cost. This method depends on cell wall compositions such as polysaccharides, which include amino, carboxyl, phosphate, and sulfate groups. According to a previous study, the biosorption method was applied to remove heavy metals using natural microorganisms [38], or as a bio-nanocomposite material, which were synthesized from microorganisms [39]. Biosorption depends on some mechanisms, namely adsorption, ion exchange, chelation, and complexation. While bioaugmentation, it is based on the metabolic capabilities of microorganisms for the detoxification of several compounds, including recalcitrant pollutants. Therefore, microorganisms can resist these toxic substances of industrial sewage through some mechanisms, among which figure biosorption, bioaccumulation, enzymatic reduction, SOS response, and enzymatic DNA repair system... etc. [40]. So, these mechanisms can be an effective way to remove the toxicity of the industrial sewage.

Furthermore, there are three approaches of bioaugmentation depending on the origin of these added microorganisms:

- Autochthonous bioaugmentation: It indicates that the microorganism is isolated from the same contaminated medium to be treated (native or indigenous microorganism).

- Allochthonous bioaugmentation: where the medium of isolation is different from the contaminated medium to be treated (endogenous microorganism).

- Gene bioaugmentation: this is when the inoculated microorganism is genetically modified to have certain functions.

Tannery effluents are known for their high salinity, due to the high use of salts during the tanning process. Therefore, [41] added a consortium of halophytic bacteria in the sequencing batch reactor to treat this tannery sewage. Despite the use of a high salt concentration of 34 g/L, the treatment achieved great abatement rates of 95%, 93%, 96%, and 92% respectively for COD, orthophosphate ions, NTK, and suspended solids.

Enterobacter sp. DU17 was isolated from the tannery effluent [42]. This bacterium was used to reduce hexavalent chromium. Indeed, the reduction rate of Cr(VI) could reach 100% when the initial chromium VI concentration is around 100 mg/L, and when glucose or fructose are carbon source. This high Cr(VI) reduction capacity by *Enterobacter sp.* DU17 has been justified by the presence of chromium reductase enzyme.

Several bacterial have shown their capacity to biosorb heavy metals such as chromium. Likewise, [11] showed that *Bacillus sp., Enterobactera erogenes,* and *Bacillus pumilus* are also chromate bacteria.

In conclusion, this biotechnology may be the most efficient and inexpensive technique for the treatment of industrial sewage because it involves the use of the most efficient microorganisms for each pollutant type.

2.2.2 Anaerobic treatments

Anaerobic treatments are generally the same as aerobic treatments but in the absence of oxygen. So, they consist of the degradation of effluents by anaerobic microorganisms. Although these anaerobic treatments have a low removal of COD and BOD_5, anaerobic co-digestion produces biogas from the organic matters. Indeed, biogas production passes through four stages under the intervention of fermentative bacteria, then acidogenic bacteria, followed by acetogenic and methanogenic bacteria [43]. The produced biogas contains a mixture of methane (50–75%), carbon dioxide (30–40%), and some traces of other components [43].

This anaerobic process can treat industrial sewage such as mill olive, agrifood, domestic, and tannery effluents [44]. According to several studies, sulfides inhibit the proliferation of methanogenic bacteria [45]. Nevertheless, a study has shown that tannery effluents can be anaerobically degraded [44]. In this study, tannery effluents were mixed with the plant of *Phragmites karka*, and then they were incubated in the SBR under anaerobic conditions using different concentrations of the plant. This co-digestion of this mixture produced 0.26 L of methane per 1 g of COD eliminated (71%), where the plant percentage was about 25%. This rate of produced biogas decreased when the concentration of the plant increases.

In conclusion, anaerobic treatment becomes very attractive due to its production of renewable energy.

2.3 Coupled treatments

Although physicochemical and biological treatments are efficient, certain limits reduce their performance. Industrial sewage is very complex and toxic, so a physicochemical or even biological process is unable to eliminate the entire pollutant load, especially inorganic pollutants. For this reason, several researchers have combined two processes or more in order to increase the purification of these effluents and to obtain an effluent that fully meets discharge standards.

In the reference [46], they coupled the chemical process of ozone oxidation with the membrane bioreactor for the treatment of tannery effluents. The coupled treatment of these two processes produced a small amount of sludge (0.03 Kg sludge/kg COD removed), which was considered to be the lowest.

On the other hand, [47] has combined chemical coagulation using ferric chloride with advanced oxidation techniques (photo-oxidation, homogeneous oxidation, and photo-fenton) for the treatment of industrial sewage, where the coagulation coupled to photo-fenton is considered the best.

In another study, aluminum sulfate and ferric chloride were used to remove organic carbon and chromium before biological treatment of tannery effluents by the SBR. This study showed that aluminum sulfate is more effective than ferric chloride in terms of COD removal [48]. However, [36] has used ferric chloride as a coagulant following by the treatment through the SBR. This combined treatment gave 99.89%, 99.98%, and 99.99% respectively for the COD, the sulfide ions, and the total chromium, and then the treated effluent was well conformed to standards.

In [49], they have coupled coagulation with activated carbon adsorption to treat industrial effluents, where lime was used as a coagulating agent. This combined treatment removed 97% of suspended solids, 99% of color and turbidity, 98% of total phosphorus, and 99.7% of chromium.

On the other hand, [50] has studied the treatment of tannery effluents by coupling 3 processes: 2 anaerobic bioreactors, followed by ozone oxidation, followed by biofiltration. The filtration is carried out under aeration into ceramic-lined column, which is inoculated with activated sludge. The optimization of this system has led to the production of a satisfactory rate of biogas and a good elimination of COD, total chromium, chromium VI, total nitrogen, and suspended solids.

3. Conclusion

The pollution generated by industries has harmful impacts on the environment and human health. In addition, their effluents were classified as very dangerous due to the presence of recalcitrant pollutants. This imposes a prior treatment of these effluents before their discharge into the environment. In this regards, different physicochemical and biological techniques for the treatment of these effluents have been shown in this chapter such as reverse osmosis, membrane filtration, oxidation, adsorption coagulation, classic activated sludge, sequencing batch reactor, membrane bioreactor, bioremediation, and anaerobic processes. Indeed, each technique has advantages but also has certain limits. For that, the choice of a treatment system is linked to numerous criteria namely the nature of the effluent, the presence of toxic substances, the operating and investigation costs and the possibility of its application at a large scale. Generally, although physicochemical techniques are very efficient and well adapt with industrial sewage, they are expensive and could generate other pollutants. Otherwise, the presence of huge amount of recalcitrant pollutants is the main limit of biological but they are also more efficient, simple, eco-friendly and especially inexpensive.

Based on this study, we considered further investigating the treatment of industrial sewage through biological processes, bioremediation techniques in particular, because they are promising, attractive and emerging technologies.

Conflict of interest

The authors declare no conflict of interest.

Author details

Karima Elkarrach[1*], Fatima Atia[1], Anass Omor[2], Omar Laidi[1], Saloua Biyada[1], Mohamed Benlmelih[1] and Mohammed Merzouki[1]

1 Laboratory of Biotechnology, Environment, Agri-food, and Health, Science Faculty of Dhar El Mahraz, University of Sidi Mohamed Ben Abdallah, Fez, Morocco

2 Laboratory of Electrochemistry Engineering, Modeling and Environment, Science Faculty of Dhar El Mahraz, University of Sidi Mohamed Ben Abdallah, Fez, Morocco

*Address all correspondence to: karima-elkarrach@outlook.com

References

[1] Naushad M. A new generation material grapheme: applications in water technology. 1st ed. Springer international publishing, 2019. 476 p. Doi: 10.1007/978-3-319-75484-0

[2] Tchounwou PB, Yedjou CG, Patlolla AK, Sutton DJ. Heavy metal toxicity and the environment. Exp Suppl. 2012; 101:133-164. Doi:10.1007/978-3-7643-8340-4_6

[3] Sharma B, Dangi AK, Shukla P. Contemporary enzyme based technologies for bioremediation: A review. Journal of environment Management. 2018;210: 10-18. DOI: 10.1016/j.jenvman.2017.12.075

[4] Junior OG, Santos MGB, Nossol ABS, Starling MCVM, Trovo AG. Decontamination and toxicity removal of an industrial effluent containing pesticides via multistage treatment: Coagulation-flocculation-settling and photo-fenton process. Process safety and environmental protection. 2021;147:674-683. https://doi.org/10.1016/j.psep.2020.12.021

[5] Asaithambi P, Govindarajan R, Yesuf MB, Selvakumar P, Alemayehu E. Investigation of direct and alternating current-electrocoagulation process for the treatment of distillery industrial effluent: Studies on operating parameters. Journal of Environmental Chemical Engineering. 2020. https://doi.org/10.1016/j.jece.2020.104811

[6] Mahto A, Aruchamy K, Meena R, Kamili M. et al. Forward osmosis for industrial effluents treatment-sustainability considerations. Separation and purification technology. 2021; 254: 117568. https://doi.org/10.1016/j.seppur.2020.117568

[7] Omor A, Rais Z, El Rhazi K, Merzouki M, El Karrach K, Elallaoui N, Taleb M. Optimization of the method wastewater treatment of unit bovine hides's unhairing liming. Journal of Materials and Environmental Sciences. 2017; 8:1235-1246

[8] Kebede TG, Mengistie AA, Dube S, Nkambule TTI, Nindi MM. Study on adsorption of some common metal ions present in industrial effluent by *Moringa stenopetala* seed powder. Journal of environment chemical engineering. 2017;6 (1): 1378-1389. https://doi.org/10.1016/j.jece.2018.01.012

[9] Boczkaj G, Fernandes A. Wastewater treatment by means of advanced oxidation processes at basic pH conditions: A review. Chemical Engineering journal. 2017;15:608-633. https://doi.org/10.1016/j.cej.2017.03.084

[10] Elkarrach K, Merzouki M, Laidi O, Biyada S, Omor A, Benlemlih M. Sequencing Batch Reactor: Inexpensive and efficient treatment for tannery effluents of fez city in morocco, Desalination Water Treat. J. 2020a;203:1-7. Doi: 10.5004/Dwt.2020.26151

[11] Elkarrach K, Merzouki M, Biyada S, Benlemlih M. Bioaugmentation process for the treatment of tannery effluents in fez, morocco: an eco-friendly treatment using novel chromate bacteria, Water Process Eng. 2020b;38:101-589. Https://Doi.Org/10.1016/J.Jwpe.2020.101598

[12] Castro L, Blazquez ML, Gonzalez F, Munoz JA, Ballester A. Biosorption of Zn(II) from industrial effluents using sugar beet pulp and F. vesiculosus: From laboratory tests to a pilot approach. Science of the total environment. 2017;15: 856-866. Doi:10.1016/j.scitotenv.2017.04.138

[13] Chandrasekhar SS, Vaishnavi D, Sahu N, Sridhar S. Design of integrated membrane bioreactor process for effective and environmantally safe

treatment of highly complex coffee industrial effluent. Journal of Water process engineering. 2020;37:101436. https://doi.org/10.1016/j.jwpe.2020.101436

[14] Sani K, Kongjan P, Pakhathirathien C, Cheirslip B, et al. Effectiveness of using two-stage anaerobic to recover bio-energy from high strength palm oil mill effluents with simultaneous treatment. Journal of Water process engineering. 2021; 39: 101661. https://doi.org/10.1016/j.jwpe.2020.101661

[15] Trishitman D, Cassano A, Basile A, Rastogi NK. Reverse osmosis for industrial wastewater treatment, Current and forward osmosiss: Principles, Applications, Advances. 2020;207-228. https://doi.org/10.1016/B978-0-12-816777-9.00009-5

[16] Ravikumar YVL., Kalyani S, Satyanarayana SV, Sridhar S. Processing of pharmaceutical effluent condensate by nanofiltration and reverse osmosis membrane techniques', Journal Of The Taiwan Institute Of Chemical Engineers. Taiwan Institute of Chemical Engineers. 2014;45(1):50Ř56. DOI: 10.1016/J.Jtice.2013.09.021

[17] Buscio V, Crespi M, Gutiérrez-Bouzán C. Sustainable dyeing of denim using indigo dye recovered with polyvinylidene difluoride ultrafiltration membranes, J. Clean. Prod. 2015;91:201-207. Doi: 10.1016/J.Jclepro.2014.12.016

[18] Gönder ZB, Arayici S, Barles H. Advanced treatment of pulp and paper mill wastewater by nanofiltration process: effects of operating conditions on membrane fouling, Separ. And purification Technol. 2011; 76(3): 292-302. DOI:10.1016/j.seppur.2010.10.018

[19] Padaki M et al. Membrane technology enhancement in oil-water separation. a review, Desalination. 2015; 357: 197Ř207. Doi: 10.1016/J.Desal.2014.11.023

[20] Miller DJ, Kasemset S, Paul DR, Freeman BD. Comparison of membrane fouling at constant flux and constant transmembrane pressure conditions. J. Membr. Sci. 2014; 454: 505-515. https://doi.org/10.1016/j.memsci.2013.12.027

[21] Basaran G, Kavak D, Dizge N, Asci Y, Solener M, Ozbey B. Comparative study of the removal of nickel (II) and chromium (VI) heavy metals from metal plating wastewater by two nanofiltration membranes. Desalination Water Treat. 2016;57(46): 1-11. https://doi.org/10.1080/19443994.2015.1127778

[22] Aziza A, Abdeljalil Z, Abdelali I. Utilisation d'un nouveau bio-floculant extrait de cactus marocain dans le traitement des rejets chargés de chrome (VI) par le procédé de coagulation floculation. Afrique SCIENCE. 2009;05(3):25 - 35

[23] Gaogui J, Shuai R, Stephen P, Wei S, Przemyslaw B, Zhiong G. Electrocoagulation for industrial wastewater treatment: an updated review, Environ.Sci.: Water res. Technol. 2021. https://doi.org/10.1039/D1EW00158B

[24] Kalra SS, Mohan S, Sinha A, Singh G. Advanced oxidation processes for treatment of textile and dye wastewater: a review. 2nd Int. Conference on Environmental Science and Development. 2011; 4: 271-275

[25] Duranceau SJ, Trupiano VM, Lowenstine M, Whidden S, Hopp J. Innovative hydrogen sulfide treatment methods : Moving beyond packed tower aeration. FLORIDA Water Resour. J. 2010;1-8

[26] Snyder EG. Elimination of odor at six major wastewater treatment plants. Water Environ. Federation, 2016;57(10):1027-1032

[27] Malwade K, Lataye D, Mhaisalkar V, et al. Adsorption of hexavalent chromium onto activated carbon derived from leucaena leucocephala waste sawdust: kinetics, equilibrium and thermodynamics, Int. J. Environ. Sci. Technol. 2016;13(9):2107-2116. Doi: 10.1007/S13762-016-1042-Z

[28] Alvarez-Torrellas S, Munoz M, Zazo JA, Casas JA, Garcia J. Synthesis of high surface area carbon adsorbents prepared from pine sawdust-onopordum acanthium L. for nonsteroidal anti-inflammatory drugs adsorption, J. Enviro. Manage. 2016;183:294Ŕ305. Doi: 10.1016/J. Jenvman.2016.08.077

[29] Zhang Y et al.,Effects of modified fly ash on mercury adsorption ability in an entrained-flow reactor, FUEL. Elsevier Ltd. 2014;128:274-280. Doi: 10.1016/J.Fuel.2014.03.009

[30] Koçer O, Acemioğlu B. Adsorption of basic green 4 from aqueous solution by olive pomace and commercial activated carbon: process design, isotherm, kinetic and thermodynamic studies', Desalination Water Treat. 2016;57(35):16653Ŕ16669. Doi: 10.1080/19443994.2015.1080194

[31] ElMouhri G, Merzouki M, Miyah Y, Elkarrach K, Mejbar F, Elmountassir R, Lahrichi A. Valorization of two biological materials in the treatment of tannery effluents by filtration, Moroccan Journal of chemistry. 2019;7(1):183-193. https://doi.org/10.48317/IMIST.PRSM/morjchem-v7i1.14064

[32] Giles CH et al. Studies in adsorption. Part xi. a system of classification of solution adsorption isotherms, and its use in diagnosis of adsorption mechanisms and in measurement of specific surface areas of solids, J. Chem. Soc. 1960;846:3973-3993. Doi: 10.1039/Jr9600003973

[33] Patil PG, Kulkarni G, Smt SV, Kore M, Shri V, Kore S. Aerobic sequencing batch reactor for wastewater treatment: a review. IJERT. 2013; 2(10)

[34] Dohare D, Kawale M. Biological treatment of wastewater using activated sludge process and sequential batch reactor process - a review, Int. J. Eng. Sci. Res. Technol. 2014;3(11)

[35] Elkarrach K, Merzouki M, Laidi O, Omor A, Biyada S, Benlemlih M, Combination of chemical and biological processes for the treatment of tannery effluent of Fez city in Morocco, Desalination and Water Treatment. 2021a; 220: 109-115. doi: 10.5004/dwt.2021.26989

[36] Elkarrach K, Merzouki M, Laidi O, Benlemlih M. Aerobic denitrification using Bacillus pulminus, Arthrobacter sp., and *Streptomyces lusitanus*: Novel isolated denitrifying bacteria. Bioresource Technology Reports. 2021b; 14: 100663. https://doi.org/10.1016/j.biteb.2021.100663

[37] Laidi O, Merzouki M, El Karrach K, Benlemlih M. Brassware wastewater treatment optimization in the city of Fez with sequencing batch reactor using activated sludge. JMES. 2015; 6(6):1562-1569

[38] Samuel MS, Abigail MEA, Chidambaram C. Isotherm modelling, kinetic study and optimization of batch parameters using response surface methodology for effective removal of Cr(VI) using fungal biomass. PLoS ONE 2015;10(3)/e0116884. Doi:10.1371/journal.pone.0116884

[39] Samuel SM, Subramaniyan V, Bhattacharya J, Chidambaram R, Qureshi T., Pradeep Singh ND. Ultrasonic-assisted synthesis of Graphene oxide – fungal hyphae: An efficient and reclaimable adsorbent for Chromium (VI) removal from aqueous solution, Ultrasonics Sonochemistry.

2018. Doi: https://doi.org/10.1016/j.ultsonch.2018.06.012

[40] Ramírez-Dírz MI, Díaz-Pérez C, Vargas E, Riveros-Rosas H, Campos-Garcéa J, Cervantes C. Mechanisms of bacterial resistance to chromium compounds. Biometals. 2008;21:321e332. DOI: 10.1007/s10534-007-9121-8

[41] Lefebvre O, Vasudevan N, Torrijos M, Thanasekaran M, Moletta R. Halophilic biological treatment of tannery soak liquor in a sequencing batch reactor. Water Res. 2005;39(8):1471-1480. https://doi.org/10.1016/j.watres.2004.12.038

[42] Rahman Z, Singh VP. Cr(VI) reduction by *Enterobacter sp.* du17 isolated from the tannery waste dump site and characterization of the bacterium and the Cr(VI) reductase . Int. Biodeterior. Biodegradation. 2014;9: 97-103. Http://Dx.Doi.Org/10.1016/J.Ibiod.2014.03.015

[43] Abbasi T, Tauseef SM, Abbasi SA. Anaerobic digestion for global warming control and energy generation-an overview. Renew. Sust. Energ. Rev. 2012;16:3228-3242. DOI: 10.1016/j.rser.2012.02.046

[44] Mekonnen A, Leta S, Njau KN. Co-Digestion of tannery wastewater and *Phragmites Karka* using a laboratory scale anaerobic sequencing batch reactor (ASBR). New York Sci. J. 2016; 9(1)

[45] Hashem MA, Nur-A-Tomal MS, Bushra SA. Oxidation-Coagulation-Filtration processes for the reduction of sulfide from the hair burning liming wastewater in tannery. J. Clean. Produc. 2016; 127, 339 Ŕ342. Available at: Http://Dx.Doi.Org/10.1016/J.Jclepro.2016.03.159

[46] Di Iaconi C, Lopez A, Ramadori R, Di Pinto AC, Passino R. Combined chemical and biological degradation of tannery wastewater by a periodic submerged filter (SBBR), Water Res. 2002;36: 2205Ŕ14. https://doi.org/10.1016/S0043-1354(01)00445-6

[47] Naumczyk J, Rusiniak M. Physicochemical and chemical purification of tannery wastewaters [Thesis]. Technical university of WARSAW of Poland, Faculty of environmental engineering; 2005

[48] Song Z, Williams CJ, Edyvean RGJ. Treatment of tannery wastewater by chemical coagulation, Desalination. 2004;164:249-259, https://doi.org/10.1016/S0011-9164(4)00193-6

[49] Ayoub GM, Hamzeh A, Semerjian L. Post treatment of tannery wastewater using lime/bittern coagulation and activated carbon adsorption. Desalination. 2011; 273:359-365. https://doi.org/10.1016/j.desal.2011.01.045

[50] Chen F, Li X, Luo Z, Jing M, Zhu Q, Zhang S. Treatment of tannery wastewater using a combined UASB (2 stage)-ozonation-BAF system. Desalination Water Treat. 2018;116:277-283. DOI:10.5004/dwt.2018.22610